T,

reading about my brothers, your fellow patriots. Thanks for your service to our country. — Best wishes Bill Knight 12-11-04

KNIGHT'S HILL

THE STORY OF A FAMILY IN WAR

BILL KNIGHT

TRAFFORD

Note for Librarians: a cataloguing record for this book that includes Dewey Decimal Classification and US Library of Congress numbers is available from the Library and Archives of Canada. The complete cataloguing record can be obtained from their online database at:
www.collectionscanada.ca/amicus/index-e.html
ISBN 1-4120-3692-5
Printed in Victoria, BC, Canada

TRAFFORD

Offices in Canada, USA, Ireland, UK and Spain
This book was published *on-demand* in cooperation with Trafford Publishing. On-demand publishing is a unique process and service of making a book available for retail sale to the public taking advantage of on-demand manufacturing and Internet marketing. On-demand publishing includes promotions, retail sales, manufacturing, order fulfilment, accounting and collecting royalties on behalf of the author.
Book sales for North America and international:
Trafford Publishing, 6E–2333 Government St.,
Victoria, BC V8T 4P4 CANADA
phone 250 383 6864 (toll-free 1 888 232 4444)
fax 250 383 6804; email to orders@trafford.com
Book sales in Europe:
Trafford Publishing (UK) Ltd., Enterprise House, Wistaston Road Business Centre,
Wistaston Road, Crewe, Cheshire CW2 7RP UNITED KINGDOM
phone 01270 251 396 (local rate 0845 230 9601)
facsimile 01270 254 983; orders.uk@trafford.com
Order online at:
www.trafford.com/robots/04-1520.html

10 9 8 7 6 5 4 3

CONTENTS

Roy and Martha Knight – 1915

Dedicated to Roy and Martha Knight and to all the heroes who have helped preserve our republic.

INTRODUCTION

MY FAMILY CREATED this story. I want to tell you about my brothers, little sister, and Mamma and Daddy. I have been trying to find a way to get the story in written form for at least fifty years. In the context in which it is written perhaps it is well I waited, for the story would have been incomplete. It is time to compile the many stories into book form. I have spent a lifetime observing, listening and reading stories of my brothers' heroic deeds. I served, but earned no medals. It is time for me to step to the plate.

This is about a family living through and contributing to four wars. Yet, superseding that theme is one of love: parental love, filial love, brotherly love, the love of God and country. We do not risk our lives without a reason. Man does not act with intrepidity to prove his physical prowess or for exhibition patriotism, but to survive and protect those around them whom they have grown to love. A patriot must be willing to sacrifice. People with ideals and character made the nation and continue to maintain it through self-sacrificing love.

The marriage of my parents, Roy Abner Knight to Martha Cordelia Holder on October 31, 1915 in the farm country of North Central Texas is the place and time that initiates the story. Roy, a handsome young man of twenty and Martha, beautiful and innocent on her eighteenth birthday were married in a simple ceremony by Rev. Bob Maddux, a Baptist Preacher. They stood by the preacher's front gate underneath a large post oak tree with leaves losing their green and beginning to fall. (Parker County Archives)

The confluence of history, necessity and the random circumstance of birth produced the first seven sons in succession followed by one daughter, the last child. They came along during World War I, the Roaring Twenties and The Great Depression. All the sons would serve in the military and the sister would have a forty-year career in the civil service, working for the Department of Defense. If we stopped here, we would have an unusual story. The actions of some of these sons and grandsons during the time of war, however, are the essence of the book. The focus of the story is on Jack, the oldest son, Daddy's right-hand man and leader of brothers and men. He will earn our nation's highest honor for valor while confronting his own mortality. My only contribution is in the telling.

The contents of this book are derived from Jack's letters and my memories, supported by other family letters, interviews and document research. I will stand by the accuracy of the undocumented events with one reservation. As I researched and discussed events with my brothers and sister I discovered that my memory of details would conflict with theirs. Therefore, my account may vary slightly from the actual fact through flawed, rearranged or added elements to my memory bank. I have to live with my memories and so will the reader.

Most of the dialogue has been reconstituted based on my partial memory and the memory of others, and by my acquaintance with the characters. I asked others in my family to read the book and evaluate the conversations for authenticity. They believe that the conversations included could have occurred. Historically, there are two events included that are not actual ones, the stories about the musical and the conversation with Mamma about heaven and hell. Instead, these are distillations of all the musicals I have attended and my conversations with Mamma and those that I overheard.

I began this work to honor my heroic brothers and nephews who faced the enemy in combat, but it is more. Great credit must go to my parents. From that day in 1915 to the present, their lives and legacy can be viewed as a microcosm of this nation's history for that period. As they began, unsure, under funded, unaware, yet following the script of their world, they dug and tugged as life unfolded in layers of hard work in the crops, sons, salvation, joy, inspiration, worry, depression, a daughter, delight, education, dread, war, despair, death, grandchildren and more joy.

They sweated, bent and strained to suck life from the abused and wasted soil of the pioneer land, two generations behind. Character was inherited and forged in this cauldron of blizzard and drought, gentle rains and tornadoes, Texas wildflowers, cool pools in the creek, fish and game, baseball and the church. The soul was battered by hardship and the guilt ever forced by human desire, temptation and submission. But the soul survived and thrived through undeniable, unqualified love for one another.

The foundation, a quarter of a century in the making, will hold up and launch a half-century of accomplishment that is entwined with the defense and enhancement of our Democracy. This is the story of one family in the America of the twentieth century.

CHAPTER 1

BROTHERS AND SONS

February 2, 1945

Jack, the oldest brother, spotted movement as two Japanese soldiers jumped to run. He quickly took a knee and fired several shots as quickly as he could pull the trigger of his carbine, killing both of them.

Jack, Curtis, the second brother, and the troopers were now under heavier bombardment from enemy artillery and mortar fire. Snipers in the trees made the advance increasingly hazardous.

After more than four years of training and waiting, 'The Boys' and their National Guard unit were now in combat. Jack had replaced his feelings of frustration and anticipation for thoughts of taking his troopers safely through the day. To lead he would be out front doing more than he expected of his troops. Curtis, who had spent a life-time following and competing with Jack would try to match him step for step, shot for shot as they faced the strongly fortified hill ahead.

November 1944

Jack and Curtis had become the top officer and top non-commissioned officer of Troop F. This came four years after 'The Boys' joined the troop as buck privates.

Mamma's anxiety was increasing exponentially. Herchel, her fourth son had already seen combat aboard the U.S.S. Maryland only half a year after being drafted. Her third son, Loyd, a third of 'The Boys' triumvirate whose National Guard unit had been called to federal service in November 1940, was in Hawaii waiting to move to the island of Pelilieu. He was a first lieutenant, serving as platoon leader in an anti-tank battery.

Mamma had taken a job at Camp Wolters near our home to help do something for our boys fighting the war. During her month of public work she met one of the mothers of a local boy who was under Jack's command. Mrs. Wilson told her in front of other neighbors how much 'Little Red' respected Jack and Curtis. Mamma was very proud of her boys.

April 2, 1967

"Major Roy A. Knight entered an area of extremely heavy and accurate antiaircraft fire in a low-level search for a downed crewman. Working at altitudes of less than two hundred feet in withering hostile fire for more than an hour, he continually exposed himself in an attempt to make visual contact with the pilot." (Citation for the Silver Star)

We suffered tragedy in World War II, but three brothers chose the military as careers. Roy, Jr., the sixth brother, joined the U.S. Air Force in 1947 and was one year away from retirement eligibility when he was sent to fight in the Viet Nam War.

May 14, 1967

"Major Knight led the rescue force through rugged mountainous terrain, poor visibility and extremely intense and accurate antiaircraft fire to the area of the downed pilot. He successfully located the downed pilot and directed the helicopters in for the rescue under extremely adverse conditions." (Citation for the Distinguished Flying Cross)

All of us were trying to get on with our lives. Roy had built a house on the family farm for his wife Patricia and the children Roy III (Chip), Gay and Bryan to live in while he was called away to duty that could be in any part of the globe. The children were anticipating summer vacation.

I had moved my family away from our hometown to take my first school superintendent position. All of us knew how dangerous his assignment was and were agonizing over his chances to make it through his tour of duty.

May 19, 1967

"Against overwhelming odds…Major Knight pressed his attack. His aircraft was struck, resulting in loss of control. Fully realizing that he could not regain control, he jettisoned his ordnance on the target in a valiant attempt to destroy it. His aircraft subsequently impacted in the target area." (Citation for the Air Force Cross)

I was completing my second year at Saltillo Schools and was preparing to leave on Monday with the graduating seniors to Carlsbad Caverns and San Antonio. We made a quick trip to Millsap to console Patricia and the children and to grieve. The senior trip was not a joyous one.

June 1944

Herchel was the first son to see combat. U.S. forces invaded Saipan with the usual bombing and bombardment and the usual air fight with the Japanese carrier based planes. Shortly thereafter, on June 22, Herchel's ship, the USS Maryland was hit. The deck log made the following notation, "1952 A Mitsubishi Type Dive Bomber appeared over the bow of the USS PENNSYLVANIA on bearing of about 030 degrees…headed for the ship and banked passing down the port side of the ship, afterwards the ship was struck by a torpedo between frames 8 and 11." (U.S.S Maryland Deck Log, World War II, Records of the Bureau of Naval Personnel (R.G. 24), National Archives at College park, Md.)

Jack and Curtis had moved with the 124th Cavalry Regiment (mounted) to Ft. Riley, Kansas. Here they turned their horses loose on the reservation and prepared to ship out to Burma. They would take in remnants of Merrill's Marauders to form another brigade in the battle against the Japanese.

1975

Jack, Herchel's son, served in Southeast Asia in 1975. He flew out of Udorn, Thailand. On the day of the evacuation of the American Embassy in Saigon, he flew as Weapon Systems Officer in a formation of four F-4 aircraft providing cover for the helicopters involved in the transport of evacuees.

Roy's status had changed from MIA to KIA. The children were growing up. Chip had completed high school. I had moved on to another job and my oldest son was in high school. The insane war was about to mercifully come to an inglorious end.

February 2, 1945

As Jack and Curtis topped the hill, the northernmost part of the Loi Kang Ridge, Jack quickly surveyed his immediate surroundings of thick scrub brush, sparser than the first leg of their approach. It was 0710 hours. Drawing no fire and seeing no evidence of Japanese he yelled to his men, "There's nothing up here! Come on up!"

It was Friday. Halfway around the world, Roy who turned fourteen the day before, June and I were in school looking forward to a free weekend playing with the Sims kids. They lived across Highway 180 about a quarter mile south. Both Mamma and Daddy were working at Camp Wolters. R.C., the fifth son had quit school, was working and thinking about joining the army.

September 14, 1997

Major Bryan Knight, son of Roy, Jr., tried, but couldn't control his aircraft. He was trying to guide it away from the thickly populated neighborhood below to a nearby lake. The F117 went nose down. His few seconds for action had passed.

My thoughts were rather dulled as I watched the news programs this late Sunday afternoon. I had retired and no longer had to face the Sunday night blues, thinking about problems at the school district I had headed.

I saw the report of the stealth fighter crashing at an air show in Bowley's Quarters, Maryland. A few minutes later, my niece, Gay, called to tell me that it was her little brother's plane that had crashed. She had heard from Bryan and he was O.K.

October 1944

Herchel's ship was one of the first to be attacked by the Kamikaze. "The USS Maryland had reached the Leyte Gulf by October 19...and on October 25,

1812.5 hours…the Maryland was hit by a Kamikaze." (Archives, Deck Log USS Maryland)

The U.S. forces were dumbfounded by this apparently insane and desperate move by the Japanese. The Maryland was dry-docked for repairs and wouldn't see action again until the invasion of Okinawa.

February 2, 1945

Curtis took cover behind a tree to search for Japanese positions. Through a clearing, he saw a two-man mortar crew running up the hill. He fired the first shots in the battle, hitting one of the soldiers. The other one grabbed his buddy and dragged him behind some cover.

Joyce and Glenda, wife and daughter, were waiting for mail from Curtis. They were living at her mother's in Poteet, Texas. He was trying extremely hard to kill Japanese, avoid death or injury while thinking about his family.

1945

Loyd finished the war on the island of Pelelieu. He spent his time there doing some patrol duty, looking for Japanese stragglers, training, playing poker and softball and waiting for the invasion of Japan. He earned three campaign ribbon bronze stars.

Loyd tried to get transferred to Troop F, 124th Cavalry after he got word of Jack and Curtis. He was denied. He was a part of the plans to invade Japan.

June 25, 1996

"Immediately after the terrorist bombing of Kobar Towers…Bryan went to Building 131, discovered a damaged medical vehicle near the crater and removed all medical supplies. With these resources he began to treat the severe wounds of several personnel. He directed ambulances as they arrived on the scene to insure that injured members received treatment. He then began a search in the rubble of the dormitory where he found three airmen buried underneath the debris." (Citation for the Air Force Commendation Medal with valor device)

Our nation had entered another era, one in which the enemy became much more difficult to identify and imminently more treacherous in its methods of warfare. Bryan, devoted to duty, would carry on the family tradition.

February 2, 1945

"Jack was hit by shrapnel from a Japanese grenade. "You little bastard!" He had sustained a non-vital flesh wound, a laceration on his forehead. Blood from the wound was running into his eyes. One eye was swollen shut. He had run out of grenades and ammo. He retreated a few steps to Lt. Leo Tynan, the artillery forward observer, for carbine ammunition and to Sergeant Bill McQuary, "Willie, give me some of your grenades."

Two days later Mamma wrote to Jack. In part, she said,

"We've been hearing news from Burma, but none the last 2 or 3 days. Sure hope you are all getting a rest.'

"Jack when you write say something about Red Wilson so I can tell his mother & if he can write tell him to write her for she wants to hear from him."

April 17, 1945

Herchel heard the gunfire changing from high caliber to lower caliber AA, meaning only one thing. A Japanese airplane was coming at the Maryland. He told his companions to dive into a depression on the deck. They made it an instant before the plane hit and exploded. The Maryland returned from Okinawa to Pearl Harbor for repairs. The war was over for the USS Maryland and Herchel.

We didn't know it at the time, but war was over for the Knight family, at least for six years. It seemed that national defense clung to our family as Curtis, Loyd and Roy became career military men. As I graduated high school, Loyd was sent to Korea. I wanted no part of a military career. I knew it could get you killed.

1951

Loyd was sent to Korea in 1951…He told Curtis that he was sitting in a foxhole and decided to move to the one next to him to get some information. Immediately after he settled into the second foxhole his first foxhole took a direct hit from a round of North Korean artillery.

May 1945

R. C., who was drafted in May 1945, was in basic training when the war was over. He later went to Japan to serve in the occupation forces.

I finally had to serve my obligatory two years in the army. R.C. and I would be the only non-combatants of the seven brothers. I joined R.C. as an occupier, serving in Germany during the last few months of occupation in 1954.

October 14, 1972

Admiral of the Fleet the Earl Mountbatten of Burma, in a speech October 14, 1972, "I told the men of the regiment, when I addressed them on 18th February 1945 that I designated the scene of this gallantry "Knight's Hill" and would have this name printed on our military maps of Burma."

"I also asked Lieutenant General Sultan, my senior American general, to quote me in support of a recommendation that Jack L. Knight be awarded the Congressional Medal of Honor posthumously." (Dedication of 'F Troop', 124th Cavalry Monument)

Curtis, Jack, Loyd

Curtis, Loyd, Jack
Ft. Bliss, 1940

'The Boys'

Jack, Curtis, Loyd   Ft. Ringgold, 1941

CHAPTER 2

BASIC TRAINING WITH HORSES

"WHERE ARE JACKCURTIS'NLOYD GOIN', DADDY?"

We had moved to the porch to watch them leave. I tried to stand as close to Daddy as possible.

"They're going to Mineral Wells to join their buddies at Camp Wolters. They'll ride a train to West Texas to be trained in the horse calvary."

"How long they gonna be gone?"

"They signed up for a year, but we'll see them before a year is up. They'll get passes to come home."

"Will they be in a war?"

"Not likely. President Roosevelt has said that our boys won't have to fight."

'The Boys', Jack, Curtis and Loyd were the oldest, twenty-three, twenty-one and eighteen respectively. They walked away from the house toward the gravel road. We stood on the porch and watched them in their army slickers through the light misting rain. They moved with a hurried step and a hint of swagger. The adventure began this November day in 1940. Three were gone. Herchel, R.C., Junior, June and I remained with Mamma and Daddy.

I had heard Daddy talk to Wade Howard and my uncles about the wars in Europe and Asia, but my mind couldn't take it all in. My seventh birthday was a month away. I had dreamed about Chinese soldiers marching down the road where I had seen the troops from the 124th Cavalry Regiment of the Texas National Guard riding on their Saturday training exercises. My brothers had joined Troop F of that regiment.

As they walked through the scattered mesquite and prickly pear on the normally arid prairie pasture past the railroad and stood by the gravel road to wait for their ride, I walked into the house. Mamma was standing by the kitchen window looking toward her boys, holding June on her hip so that she could see out. There were no tears, but a soundless, anguished cry filled the room. Those three soldiers were still her babies. That look of quiet stoicism would become her persona for the remaining thirty years of her life.

"Mamma, what's a matter?"

"Billy Rowe, I'm afraid."

"What about? They might get hurt?"

"Yes, but that's not all. If we get in a war, they could get hurt bad and they might have to kill somebody."

"Wouldn't they just kill bad people?"

"Killing is killing and the Bible says, 'Thou shalt not kill.' We're going to have to pray that the boys will be o.k. and they won't have to kill anybody."

"They'll be alright." Unspoken was my sincere belief that Mamma's will and prayers would make it so.

We stood in silence for a while. The Boys were still waiting for their ride and were becoming restless.

"I wish Gat would come on. I'm ready to show those alley rats from Mineral Wells how to ride and see if they know how to scuffle."

"Buzz, you better watch out. Some of those alley rats might be mean as hell,"

Curtis answered.

"Shit! Pot Guts, I could whip those city boys with one hand tied behind me."

Jack took this in and said, "Loyd, you'd better watch out. You don't want to get in trouble the first week. We need to be good soldiers."

Slightly taken down, Loyd muttered, "I can whip their asses and still be a good soldier."

"I know I can't tell you anything. Just be careful."

After more small talk about their expectations for their new adventure, their ride came and they were transported to a new world, one with arbitrary and mystifying rules, seemingly useless activity, and interminable waiting. The 124th Cavalry Regiment was a part of the Texas National Guard with units scattered throughout the state. The 124th was formed in March, 1929 with troops in Houston, San Antonio, Fort Worth, Corpus Christi, Mineral Wells, Brenham and Seguin. Until 1940 its most notable action was helping restore order in the East Texas Oil Field soon after its discovery in 1930. It was now being mobilized for federal service in a belated effort to prepare a defense in an uncertain and dangerous world created by the megalomaniac in Germany and the power-crazed imperialist warlords of Japan. They expected to serve on active duty for a year along with thousands of others in similar units until the United States could catch up with its regular army. (Marsmen in Burma, John Randolph, Gulf Publishing Co. and Curators of the University of Missouri, 1946, 1977.)

The victorious allies of World War I had unwittingly made a critical mistake with the Treaty of Versailles. Vengeance explains it but doesn't justify its lack of wisdom. With the flawless insight of hindsight we see that now. Nevertheless, the errors were compounded and enhanced by our willingness to wallow in this mess during the self-indulgent twenties. By the thirties the odor clung to the world as good people in a state of desensitized apathy stood by and watched the German Nazi party and Hitler's quest for Lebensraum and the Japanese warlords' ravenous expansion push the world inexorably toward a full-blown schizophrenic episode of apocalyptic disaster.

By the time The Boys joined the cavalry, Hitler had walked into Austria and the Sudetenland. His Blitzkrieg had contemptuously swept aside the proud Polish defenses led by its crack horse cavalry divisions. His quick work of bypassing the Maginot Line through the Netherlands and his rapid defeat of the great French army and the British Expeditionary Force was already history.

Beginning in 1931, the Japanese had marched through Manchuria and later proceeded to rape and plunder large chunks of China with growing savagery. Their lust for additional territory was whetted and was growing. The Battle of Britain in the summer of 1940 and into the fall was one of the most difficult and courageous efforts by a whole people, both civilian and military, ever achieved. How could our country, so morbidly isolationist, reasonably expect to escape the rapidly advancing insanity? (New History of World War II, C.H. Sulzberger with revision by Stephen E. Ambrose, American Heritage, Viking Penguin, 1966, 1994, 1997.)

The first duty of The Boys and their compatriots in Troop F was to pack for the train trip to Fort Bliss near El Paso for training. On November 28, 1940 they boarded a troop train in Mineral Wells. The train would travel east on a spur line twenty miles to Weatherford and be switched out to the Texas and Pacific rail line to travel west to their destination.

The spur line to Weatherford passed through Garner about two hundred yards north of the school that we attended. It was a school day and the four of us who were in school were expecting them to come through Garner. We were in recess at the time we heard the train coming. The baseball field that doubled as a playground was adjacent to the tracks.

Our teachers and principal may have known the time for the train's passing or perhaps sheer luck prevailed because we were at the playground. The train moved very slowly at this place. It wasn't uncommon for the local daredevils to run along the rails, grab hold of the ladders on the cars and ride for short distances for entertainment. Some would ride all the way to Weatherford fifteen miles to the east.

As we heard the train, we ran to the boundary of the school property next to the right-of-way. My teacher was nearby. Her name was Clementine Trammell.

"Miss Trammell, my brothers are on the train."

"O.K. Billy, try to spot them and wave."

Miss Trammell had been a classmate of Jack and Curtis at Weatherford College. She was pretty and I couldn't understand why one of them hadn't married her. My memory has dimmed, but I think I suggested this possibility to her. Six-year old kids from the backwoods do not possess a great deal of social poise.

We waved and yelled. The soldiers waved and yelled. My brothers were the only Garner boys who joined Troop F. Several had joined the National Guard artillery battery at Weatherford, the Parker County seat of government. I have no graphic memory of seeing my brothers, but I knew they were on the train going to an exotic place called El Paso. The sadness and fear associated with the unknown danger of them becoming soldiers and going away soon dampened my fervor at being the center of attention.

I had had my own experience of leaving home and I was just as apprehensive as I was at their leaving. In 1939, Mamma and Daddy decided to start me in school at the age of five, almost four months before I turned six. We didn't have kindergarten, so I would be in first grade. I suppose that by the seventh time the thrill or poignancy of taking a child to his first day of school had worn thin. Herchel, the oldest boy in school at the time, was assigned the task of enrolling me in school. He took me to the door of the 'dog house' that housed first and second grades and told Miss Velma Morris my name and other vital statistics and left. I had no problem because I had three brothers close by and Miss Velma was so sweet and smelled so good that I was immediately smitten. I noticed that some girls would hug and kiss her goodbye as we left to catch the bus to go home. I decided that I would try this. I succeeded and enjoyed several weeks of this intimacy before some of my rascalian peers discovered my 'sissy' conduct. Their teasing caused a reluctant cessation to this practice. What a price to pay to keep my dignity and standing.

We received a post card from Jack dated November 29, 1940 mailed from Ft. Bliss which was the correspondence that initiated an unrelenting habit of writing home an average of once a week for four years. The card also established an unwavering theme of convincing us at home that everything was o.k. and that life in the army was tolerable, if not good. There was always the implication that the Knight boys could easily take anything dished out to them. He tried, without success, to eliminate reasons for Mamma to worry. He wrote,

"Dear Bunch,

"How is everything? We are at Sierra Blanca and are getting up Thurs. morn. We are o.k. Having a big time. We are on Pullman cars and the beds are soft as mush. This town is 72 miles from El Paso. We are stopped right across a Highway & have had a truck blocked for 30 minutes. This is a pretty country; rolling hills & flat country between."

"Be good.

"Love, Boys"

The mush was a reference to one of our food staples through the depression that might be served at any meal, but usually for breakfast. It was nothing more than boiled corn meal served hot with butter and sugar or honey. It was palatable and sustaining.

I am sure that a few of Jack's letters were not kept. The first letter that survived was written two weeks later. This letter also reveals recurring topics dealing with making and saving money, encouraging the kids at home to make good grades and giving advice to all of us, including Mamma and Daddy. Jack very often reverted to colloquialism while writing. He had some problem spelling some words, but his letters were always informative, entertaining and at times prophetic. He wrote Monday, December 16, 1940 from Ft. Bliss,

"Dear Bunch,

"How is everybody? We are all fine and we do see each other 100 times a day. We've all been on details now & then but we haven't been on to much. We're learning how to aim & sight a rifle & how to pitch a pup tent.'

"They are saying around here that we are going to Ft. Clark between Del Rio & Eagle Pass. I think it's a long way from town & if it is we can just stay around camp & save our money. There is sure some good hunting & fishing country around there. I think it's a lie but it's out all over camp that we are gonna ride our horses down there. If we do it'll be just like trail driving days only we won't have any cows to drive. Boy I can't wait. I'll bet it's a lie tho.'

"It's pretty cold out here tonight but not as cold as it gets at home. You said in Buzzard's letter that it was cold & rainy. Well it has been cold at night but in the daytime you can get on the south side of a building & it is really warm. The sun just burns. I'll bet it really gets hot out here in the summer time.'

"I went to Jaurez the other night. It's as bad as the worse part of El Paso. You can't tell what in the heck they are talking about. There are a gillion little places to buy souvenirs. You have to pay two cents to cross the bridge & one cent to come back across. They just try to get your money, and it don't make them any difference how they get it. They can't get blood out of a turnip tho. I didn't take but $1.50 with me & I buttoned it up in my shirt pocket. Loyd and Curtis haven't been to over there yet but I guess they will go over before we leave. Heck I wouldn't have missed it for anything. We can't wear any part of our uniform over there.'

"We have a little norther tonight but the wind never has blown as hard here as it does all time at home. Just a light breeze the most. I was surprised this camp is built right in a sand bed. Just as sandy as that Garner sand, but is rocky all west & north of here.'

"We're fixed up good now with lights & gas and a new bath house.'

"You kids pour it on & make good grades. We'll pay you for them. That dumb Nig couldn't break anybody. Tell June this letter is half to her. Tell Buck to hold everything down around that new theatre. Boy I hope Dad can get on as a carpenter. Dad I know you can cause these guys building these barns & things don't do anything but saw & nail. The boss tells them how. You all be good."

"Love, Jack"

His obsession with money can be explained by an incident that happened the previous year that highlights the value of a dime at that time in the late stages of the Great Depression. Daddy had sent me to the mailbox with instructions to buy some stamps from the mail carrier. He gave me a dime and told me to put it in my pocket and not play with it. Never having had a dime to touch, I couldn't resist pulling it out and trying my hand at flipping it as I had seen my brothers do. Of course I dropped it into the loose sand in the roadway and it sank into the sand. I looked for that dime for a long time because I had a deep dread of going back to tell Daddy what I had done. Not only had I not minded him, I had perhaps lost the only dime that he had. However, I faced the music and just as I suspected, he taught me a lesson in his usual way. Daddy was plowing the garden with a Georgia stock, powered by a mule. He took the plow lines that he had in his hands and whacked me on the bottom. Later, he gave Curtis the dime if he could find it, to buy a loaf of bread. Curtis was in college at the time. All of us had been trained to take care of our money. Curtis got his loaf of bread.

Their life in basic training was much like my own in the summer of 1954 at Ft. Bliss in the same area of the post known as Logan's Heights. They were billeted in small tents. I was placed in wooden huts about 15' X 15' that held four men on army cots with footlockers at the end. These were carryovers from World War II days. The cracks in the floor made it possible to thoroughly sweep the floor from back to the front, arriving at the door with no sand or dust.

In Curtis' letter of December 18, 1940 he spoke of their breaking horses for remounts.

"We put saddles on them after snubbing them to a post. They buck when we put the saddles on them they are so wild. We soon learned how to catch up with short rein and hold their heads up so they won't throw us off. We will have to ride them about a month before they are trained. Rode one today named Gravy. He pitched more than any I've rode yet. Buzz and I haven't been bucked off yet. John's horse stopped to quick and fell down today. He jumped off before the horse fell. It is just like a rodeo every day, and we take part in most of them." If the horses became too unruly, they would grab them by the neck and take an ear in their mouth and bite down very hard. This technique, called 'earing down', would subdue them every time.

The only difference from my training that is obvious was that they had horses. This was a mixed blessing. They could ride on long marches. I had to walk. They had to care for themselves and give even more attention to the care of their mounts. I had only myself to care for. I believe I had the better deal.

Basic training was a departure from their last four years; college life, taking any job for a little money and the insecurity that prevailed. The college boys would come to the farm to demonstrate their newly acquired sophistication. They taught us to play touch football. Baseball, softball and basketball had been our only competitive sports experience. At my age I had only watched anyway. We had no radio, so when they taught us to sing 'Alexander's Ragtime Band' I thought I had entered a new world. You have to admit that's a departure from 'The Old Rugged Cross' or 'You are My Sunshine.'

The Boys were getting the full dose of being the lowest on the pecking order, but at the same time were looking for ways to improve their financial standing. Jack wrote on December 24,

"I'm on K.P. today. It's Xmas eve. but I'm glad cause I'll be off tomorrow & a lot of the boys have got to serve on guard & stables and kitchen but none of us will because we are all on something today. Curtis is on stable guard & Buzz is on tent guard. I think we'll go to town tonight.'

"We got your box Sat. & was sure glad. I think about 10 ate out of it. I've still got a piece of that fruit cake. Boy it was good but the other was too.'

"I guess you all think we didn't think of you Xmas but we couldn't do anything about it. We have all our money loaned out & they told us we would get paid the 20th but we didn't so we are broke. I've got 20 dollars loaned out & I'll get $25. I might decide to draw my money out of the bank and loan it. I can really make the money & it's a pretty safe risk. They will pay 25% for a month & if they want it bad enough they will pay $1.50 for a dollar." [That would make the APR either 300% or 600%.]

Jack wrote on January 7, 1941 with more details of their training.

"Our new colonal is putting us on the line now. We are getting up & having exercise & cleaning up the barracks & beds & eating breakfast before daylight. Just as soon as daylight comes we pick up match stims & stuff around the buildings & go to the stables. We have to clean up all the stables & put down bedding hay & feed alfalfa before we saddle & pack out. We are ready to go by eight o'clock with a good half a days work under our belts. We ride till 10:30 & come in & clean & soap our saddles & groom our muddy horses. In the eve. [afternoon] we foot drill two hours & have study periods in our hand books for two hours & then get ready for retreat at 5:30. We are so busy we can't do anything but work. We don't stop to go get a drink here. They don't believe in that."

By the fourth week of training the F Troopers were beginning their weapons training. Jack tells us about it in his January 8, 1941 letter.

... "We are going way up north of camp everyday now shooting our rifles. Boy we've got some good shots in here. I'm shooting pretty good. One time I got 7 bulls eyes out of 10 shots. That's an 8 in target at 2 hundred yards. Another time I got 5 for 5 on rapid fire. That was on kneeling. On standing I didn't do so good. I got 2 bulls & 7 in the 2nd circle, I missed one. I think I'll make it ok. Curtis is keeping our scores. But he stands behind us, not the targets. They talk by field telephone to some guys in a hole behind the targets. Buzz is doing good too. I don't know his scores but he's better than average. I saw him get 3 bulls in a row the other day."

Jack updates us again the next week on January 13.

"How is everybody. We are all getting along fine. None of us have been sick yet. We are still shooting. I am learning to shoot a machine gun. Boy it is lots of fun. When we got done today we carried off about a tub full of hulls. That was what about 20 of us shot up. We sure do waste lots of money that way. We just turn it loose & let it sputter. We have four guns in our troop. Our sergeant, Owens, said he was going to get me a job as a gunner, because I am a good shot. I am pretty good. If I do I'll get more money. I think Curtis is going to start cooking about next week. The mess sergeant said he could have the job. We are getting along. O yeah. Buzz made one 49 out of 50 in rifle. The best I could do was 47. He got 9 bulls & one 4 out of 10 shots. The 4 is in the second ring. A bulls eye counts 5. Only one guy beat him he made a perfect score, 10 bulls. That's on 200 yards at rapid fire. We are not going to Ft. Clark as we had thought. We are gong to Ringold. It is about 40 or 50 miles from Brownsville right in the heart of the valley. I'll bet we get fat on oranges & grapefruit next summer. It's a long way from here. We are supposed to leave here the 5th of next mo. get there on the 6th. It will be warm enough to play ball all time down there."

In their fifth week of training they were firing their weapons for record. They were trained to fire the bolt-action 1903 Springfield, the mainstay of World War I. The Garrand M-1 30.06 rifle was in production, but had not reached all the training facilities. Their ability and luck determined their rating as expert riflemen, sharpshooter, or merely marksman. As they lay prone, kneeled and stood to fire their rifles, their aim could be affected by wind gusts and blowing sand. The morning was cold and clear as they assembled for morning roll call before marching five miles north to the rifle range.

Loyd spoke, "John, I'll bet you a dollar that I qualify as an expert today."

Jack answered, "I don't bet to lose. I saw you shoot last week. I wouldn't bet against Pot Guts either. I'll bet you a dollar he beats you."

"No deal. I don't bet to lose either."

Even after a five mile hike, they weren't tired, but were anxious to get started firing. Sergeants Sam Whatley and Truman 'T' Owens were barking orders to get in firing positions, two men to a station. The targets were two hundred yards to the west with the foothills of the Franklin Mountains as a backdrop. The sun was at their backs; excellent conditions for record firing.

The Knight boys were separated and paired with other troopers. It was a good day on the rifle range for the squirrel shooting boys from Garner. Jack told the story in his letter of January 19, 1941.

… "We all shot rifles again today. Buzzard and I are shooting for other guys that couldn't hit the board much less a bull's eye. Curtis shot for Curtis & really did shoot. 202 is expert score. I made 204 for Elmore Warren. Buzz made 203 for Fred Harrison. Curtis made 217 for himself. My score was 198 & I made expert for that dern Warren & won't get any credit for it. I'm glad I did tho. I thought I could all time. I was glad to get the chance. I guess we'll all three shoot for someone else tomorrow."

If they had waited for everyone to actually qualify for themselves, they could have been there weeks longer. The non-coms would look the other way and become accomplices in this conspiracy. I assume they did it the same way we did fourteen years later. In the extreme, we would just pencil in the scores.

Basic training and their stay at Fort Bliss were about to come to an end. Jack wrote his last letter from Ft. Bliss on January 31, 1941.

... "Well we are gonna be on our way about Monday. That is if we don't have to stay here because one in the troop has measles. We can't even go to the canteen. We might have to stay here a week or two yet, dern. When we do leave we are going by Clovis, Lubbock, B-Wood, Temple, Houston, Santone & South. We are gonna play war & pretend all the railroads in South Texas are blown up. How is Grannie & H.D. getting along. I hope they are well. I guess the others are o.k.'

"Tell June that was really a sweet letter she wrote. Tell all the other kids to write to us.'

"Dad are you still working? Will this filter [for Mineral Wells water supply] be as large as the other one? Are groceries going up in M.W. I'd like to be there & see her grow. I'm not doing so bad. I made around $80. all together this mo. I'll add some every mo. as long as I can get 50%. I might have to cut down to 25% when we get down yonder. They can't spend their money like they do here. Well if they want it I've got it and if they don't I can keep it. I'm a hard hearted critter. They all look alike.'

"I imagine that when Hitler's boys take their report back on what Uncle Sam is doing over here that he will first take down his sign. Roosevelt seems to enjoy getting a chance to defy him. Well I'm thinking this war will be over by Sept. and we'll never get a chance to use our nice big army camps. I hope so anyway.'

"There is a little Italian in our tent from Wink. He knows Bob & Elton [Estes, our cousins]. He said he thought I was one of them when he first saw me. He was drafted into the cavalry. He don't know the front from the back of a horse. He calls a spur a stirrup. He just won $6.50 in a crap game & is tickeled to death. He'll probably loose it before morning.'...

Meanwhile, on the home front, boys were being boys. With Daddy working, the four male mischief-makers took advantage of the reduced supervision to broaden their repertoire of adventures. Some of the activities chosen were simply stupid. We had dug two craters in the back yard under a huge oak tree and covered them with logs and dirt to form bunkers. Herchel and R. C. decided that this wasn't enough challenge. They wanted a cave for their hide-away. It was decided that the earthen tank dam would be the perfect place. We worked all day creating a tunnel into the dam in a horseshoe shape with two entrances. Unfortunately for us the project was in plain view of the lane. Daddy saw this testimony to adolescent brain paralysis as he came home from work. I don't remember the punishment delivered, but I do remember Daddy's apoplectic tirade. I think that my observation of these events helped me develop a finer sense of prudence that helped me avoid Daddy's wrath during my teen years. Maybe it was a matter of Daddy being worn out from too much herding.

Troop F completed basic training and boarded a train for departure shortly before midnight, February 2, 1941. This calendar date would become extremely significant four years later. Their trip to their new station assignment doubled as a war training exercise as Jack pointed out. Jack was trying to write a letter to us at home, when Curtis and his buddy, Jack Stockstill approached with bulging pockets.

"Hey John, you want a bar of candy? We've got Baby Ruth, Butterfingers and Hershey's, both kinds."

"How much do you thieves charge?

"It's only a dime a bar."

"A dime a bar! You think I'm rich?"

"You may not be rich, but I know you've got a pocket full of money."

"I won't have it long at these prices. Yeah, I want two Hershey's, one of each kind. Murphy you better stop bootlegging candy. You know what Captain Dews said about that. He'll put your ass on K.P. all the way to Ft. Ringgold if he catches you."

"I don't care. I'm making money. I made five dollars last week during the measles quarantine. I'll just do my extra duty and pocket the money. I'm not losing it in a crap game like some of these suckers."

"O.K., it's your problem."

Jack returned to his letter and related the event to Mamma and Daddy. They were buying a box of twenty-four bars and selling them for ten cents each. This entrepreneurial pursuit brought them a profit of $1.70 per box. In Jack's words they were 'getting rich.'

This is an appropriate place to tell my own story of opportunism. At about this time I had an unexpected opportunity to exercise my business skills. The school put the lower elementary kids on a bus and hauled us on a rare excursion to Fort Worth to visit the Forest Park Zoo. Daddy gave me a quarter to spend. It lasted most of the day. A short time before we were to leave I had six cents left, not from thrift, but more from indecision over how to spend it. Another boy, the name long since forgotten, ran out of money before he ran out to things to buy. He offered me his pocket- knife for my six cents. I quickly made the deal for my first pocketknife, a staple for farmers and their adolescent sons. I wasn't sure Daddy would let me keep it, but I couldn't pass up a bargain like that. I was able to convince Daddy that I would be careful and he decided to trust me with it. I kept the knife for several months before it slipped out of my loose fitting overalls pocket and entered the mysterious life of the lost and never found.

The troop train traveled through Midland, Sweetwater and Brownwood before taking a southeasterly course through Brenham, the home of E Troop and Blue Bell Ice Cream. Brenham was the location of a predominantly German settlement in South Central Texas. The train stopped and was greeted by a large, cheering crowd. All the troopers were given an ice cream bar. As Loyd was licking the stick of the last clinging bits of ice cream he gazed out the window and said, "You know, these squareheads ain't all bad." This was the first crack in the wall of prejudice against the German-American troopers. Four years later Troop F would find in them a formidable companion as they battled the Japanese for control of the Burma Road.

The final leg of the journey went through Houston and Alvin to the Rio Grande River near the gulf coast at Brownsville and up the river to Rio Grande City, through the heart of the Rio Grande Valley. The tropical setting and the thousands of acres of citrus groves and vegetable fields awed The Boys. They had never seen orange trees and vast fields of vegetables. The contrast between this corporate approach to farming and the Knight family subsistence farming was gripping. As they traveled west/northwest, the valley tapered off and the orchards and farms gave way to mesquite and sage- brush, cactus and other desert flora. Farms became scarce. The Boys would not be located amongst groves of citrus fruit as they thought.

Curtis

Jack at Ft. Brown

Loyd

Troop F
Second Squadron

Parade in Rio Grande City

CHAPTER 3

FORT RINGGOLD AND RIO GRANDE CITY

BY LATE WINTER of 1941 Germany had added the Scandinavian countries to their growing sphere of influence and had occupied much of it. Japan had come into political conflict with a country with vastly more resources and capability to wage war. Their idiotic aggressiveness caused them to lose trade with the United States, thus further restricting their ability to wage war. Their solution to this dilemma was an even more suicidal plan to expand into mineral rich Southeast Asia through the Philippines. This action would bring even more enmity from the United States. The militarist leaders knew this and decided to buy time by destroying the U.S. Pacific Fleet at Pearl Harbor. Blind power ignores reason and gives birth to self-destruction. The momentum of the pernicious alliance between Germany and Japan had become irreversible and events of 1941 would bring the utter mayhem of World War II. The British sank the Bismarck on May 26. (New History of World War II, Sulzberger and Ambrose, American Heritage, Viking Penguin)

Mamma and Daddy had lived through hard times raising a large family on small farms around northwest Parker County. Nothing had prepared them for coming events. They were from large, pioneering farm families where both physical and mental toughness was required and unconditionally accepted as the standard for survival. This land of rocks and sandy loam reluctantly yielded the crops. Hard work, faith in God and unrelenting will were the common elements of those who successfully scratched a living from this unwilling land that was only intermittently blessed with rain.

With the courage of their parents and the optimism of youth Daddy and Mamma began a life together that would produce an uncommon family history. Seven sons in succession followed by an only daughter during the period from May 29, 1917 to August 29, 1936, and their lives, was Mamma's and Daddy's legacy. The five oldest brothers came of age during the late thirties and The War. They were toughened by living through two decades, working together with Daddy and Mamma doing the farming and helping in other ways for the family's welfare. Their inherent athletic ability was enhanced through playing baseball, softball and basketball. They also inherited intelligence and grit, which was enhanced by practicing their chosen religion.

At the same time, my parents knew the value of an education. Although neither of them received any high school education, their children would. Against extreme hardship, their oldest sons would attend Weatherford College during the late years of the Depression. The Methodist Church, the family's traditional church of choice, operated the junior college. Jack, at the age of nineteen, started college September, 1936 within a month of the birth of his cherished sister, June.

By 1940, with world war imminent through ambitions of mad men, the future of this close-knit family was being forged. Without consciously formulating a plan, Roy and Martha unwittingly contributed their physical, mental and religious souls to the nation's experiment in the fusion of democracy, free enterprise and personal freedom to make it a working and robust reality, even with the recent crippling effects of an economic depression.

Their union would produce individuals who, through happenstance, politics, economics, and personal choice, would have the opportunity and will to perform some uncommon feats of courage in the defense of their country. Their individual acts are not ordinary and taken together would create an unusual tale of a bunch of ordinary farm-raised brothers fighting for their country, their home and their friends.

Perhaps some of the stories are apocryphal. Others are truth even more incredulous than fiction. But they are all a part of the accepted family lore and we accept them as our history. Jack is my brother, the oldest of seven sons. I am the seventh son. Much of the story that I will tell, I am convinced, is in some way influenced by him because his spiritual presence was pervasive at the time and continues to be omnipresent today. Jack is one of those uncommon men whose life was raised by events and conditions to show intrepidity in its purest form.

Why was he this way? Perhaps it was because of his training within his large family where personal sacrifice was a way of life. A part of it may have been the physical environment of a rough, sometimes infertile, dry, unyielding land that he helped fight. Perhaps it was the absorption of the pioneer spirit that still lingered in the North Texas countryside of his youth. Anyone who could make a living from this land had to be tough. I am convinced that the greatest influence was the firm and loving parents who guided all of us to adulthood and passed to us their traits and values. Or could it be that he was one of those rare and special people that fate touches with an indomitable spirit. Whatever the reason, he was made a god of my youth and a lasting hero of my life.

The Boys were now nearing the end of their journey from Ft. Bliss to their new home at Ft. Ringgold. They were in for another culture shock as they arrived. The fort was located three blocks from Rio Grande City. It was an oasis and the inhabitants of the surrounding country were predominantly of Mexican heritage. The fort was a vestige of the western frontier of the nineteenth century, having served as a base for Indian fighters and the troops that helped protect the border from bandits, both Mexican and Anglo. Robert E. Lee and Stonewall Jackson had served there. The fort compound contained citrus and palm trees with green, manicured lawns. This was quite a contrast to the dry sand and rock at the training facility at Fort Bliss and the scraped, packed dirt yards back home. The troops were herded onto the quadrangle located in front of the barracks on the perimeter. As they stood amongst their personal gear and inspected their surroundings while awaiting formation a few subdued comments could be heard.

"What do you think of this? I believe I can handle this duty after that dry hole at Bliss."

"Yeah, it looks good now, but a die-hard pessimist like me can't really believe it."

"I bet before we're here a week they'll have us digging up the grass and hauling in rocks."

"Even if the place looks good you can bet your ass duty won't be a picnic and we'll forget all about the nice lawns and buildings."

In truth, the physical environment didn't change and the training schedule quickly dispelled thoughts of paradise. Of course, the ingenuity and irrepressible spirit of young American soldiers would find and invent ways of amusement. They wasted no time in getting sports activities started. Being genetically predisposed to athletics and having played basketball at Weatherford College and softball and baseball in amateur leagues around Parker County and Mineral Wells, the Knight boys were eager instigators and participants in the inter-troop contests. They joined force with others to form a basketball team. Although the Knight boys basketball experience gave them an edge, Jack mentioned Mickey Crossland making a significant contribution and Bob Harrison who was better than any of them and scored most of the points. Jack wrote about it in his letter of February 12, 1941.

"We played two ball games this week. We beat G Troop 66 to 4 or 6, I don't know which. Bob Harrison made 30 points and Buzz made 22 & Murphy made the rest. Me & Mickey Crosland didn't make a point. [Their prowess was on defense.] We got a little better competition from the city team here in Rio Grande. We beat them 27 to 12. Bob made 19 of them. He could make anybody's ball team."

On another subject, he added, "We are going on an over-night ride Friday night. I'll be glad when we can get out & do something. We don't do anything but ride all day now, but it is just around camp here. We are going to ride out this afternoon with full packs, guns & all to learn what we are going to have to do Friday."

Curtis wrote at about the same time, "Went to the show last night with Jack & Loyd. Saw Dorothy Lamore in Moon over Burma. Good show." He also wrote about the apparent suicide of one of their sergeants who had been despondent about personal problems.

The troopers settled into a regimen of weapons training, riding, short military exercises, sports, beer -drinking, shooting craps, playing poker, fighting E Troopers and attending chapel. It wasn't a bad way to spend an obligatory year in the army, if you could dismiss the pervasive presence of suppressive rules that are necessary to keep a collection of free-spirited citizen-soldiers in a semblance of discipline.

Loyd seemed to be a lightening rod for misadventure and high jinx. Jack talked about him in his February 20 letter.

"Dear Mom, Pop and Kids,

"How are you all? We are all o.k. Curtis & I are both on main guard today. I'm on the west gate & he's on a post that goes around the barracks & stables. We will get off this eve. at 5:00. We are on 3 hours and off 3.'

"Well it's really spring down here now. It seems like about April to me. The weeds & grass are about like they are up there in April. It's cloudy today but it has really been hot some evenings.'

"We made our ride o.k. It was just about like going down to Duke Mt. fox hunting. It was 13 miles from the post & we rode it in about 2 ½ hrs. It was pretty hard on some of them. We rode at a belly busting trot nearly all the way.'

"Boy talk about coyotes, they are really around here. They sounded just like they used to up at Jimmie's in Young County. They were on both sides of the river. There are hills around here just about like around home. The river runs right along the side of camp. It's not 200 yards from here to the river. We go down there on Sunday or whenever we want to. There's nothing on the other side. One day we saw a Mexican watering a donkey & another one plowing. That's all."

"Quite a few people come by from other states & drive through & ask questions. They all want to know how many men we have. They seem surprised when we say only about 400 & then we tell 'em we're pretty tough. They act like they think we are. I guess they want us to keep those Mexicans on their side of the creek. These people don't think any more about the Rio Grande than we do the Brazos at home. They are Mexicans on both sides anyway.'

"Bill you can't write as good as June. I believe you are just kinda goofy. Well maybe you got a little bity bit of book learning. You better not call me John any more & clean your nose or you never will get you a gal. Oh I guess you've all ready got one tho.'

"Buzz actually got a haircut. He went to the show the other night & let a guy try to hypmotize him. Nearly all of F Troop was there & they sure have been teasing him. He said him & that guy was just putting on a show. He's got a new name. It's Scaffle. He calls Red Wilson Shorty & Red started calling him that. We all weigh about the same, about 185.'

"Well I've got to write Grannie so I'll close."

"Write soon

"Love Jack"

Back: Curtis, Jack, Loyd
Front: Bill, R.C., June, Herchel, Roy Jr.
1938
Back: Murphy or Pot Guts, John, Buzzard
Front: Gopher, Smutter or Jew, June Bug, Tuffy or Bape, Nig

Not being on a wartime status, security was not severe. Later, the information they gave to civilians would be forbidden because it would indicate that only a squadron of men was stationed there. A cavalry squadron was equivalent to an infantry battalion, although with fewer men. Certainly the nickname Scaffle was intended to be Scaffold, in reference to his height. The next week brought bad news from home. Jack made reference to it in his February 26 letter.

"We have talked it over with the captain & he said we could get off but it would take us until Thursday night to get there. I'm afraid we would be late for the funeral. It's a long way & we are all broke (being two days from payday).'

"Tell all of them that we hate it as much as any of them but it seems impossible to make it.'

"Mom don't think too much about it. You know we've all got to die sometime & just do like you always told us..just be ready. I'm sure Grannie was if anybody ever was."

Jack was referring to the death of Grannie Holder. She lived in a new house built by her youngest son, H.D. at the back of the farm, near the water well. A problem was created for the remaining brothers and sisters. Vercia was the youngest daughter and mentally slow. She had only attended a few days of school before being withdrawn by her parents. Mamma and her siblings would be faced with the problem of her care for the next four years until they decided to place Vercia in a state school in Austin.

Jack mentioned his money loaning enterprise in his March 2 letter. "I didn't collect all of my money this time. Its all good tho. Some of them just paid the interest. I loan it one for one & a half or two for three & when they want more I let them have it 7 for 10 & 14 for 20." If he didn't invent the loan shark business, he surely elevated it.

On March 23, 1941, Jack wrote about some of their foolishness.

"I stay in one building with the machine gun platoon & Curtis and Loyd stay in another with the riflemen. Buzzard & J.C. Byrom, one of his buddies came down to our barracks last night & swapped our beds & lockers & 'short sheeted' us. They double the bottom sheet back & it looks just like it did when you left it, but when you start to get in it, its just like trying to crawl in a short sack. We'll make them beg tho before we're through. I'm going go put a box of Post Toasties in their bed."

Herchel was seriously ill in the spring of 1941 with pneumonia. Jack wrote about it on April 17.

"How are you all? Is Tuffy getting any better? If you can I wish you would write everyday till he gets better. You have so far I think. I hope you get along o.k. Bape. I guess you've just got a little case of the flu. Mamma don't worry about him. You know how tough he is. You couldn't get him down with a chene gun and a sledgehammer. Buzz got here o.k. He really had a good time at home. I guess he's deeper in love than ever. He ought to marry that little hag & bring her down here."

Nicknames were popular in our family. All of us had at least one and, as we see some of us had more than one. Most of the nicknames came from insults that stuck. Mamma never used those names, just us, our friends and occasionally by Daddy. Jack had only one and it wasn't one of the nasty variety because none of us were brave or foolhardy enough to pin a really bad one on him. His nickname was 'John', not from its historical association with his name, but from the name of his eccentric landlord while at Weatherford College.

Curtis was 'Pot Guts' or 'Pot-Gutted Murphy' or just 'Murphy' for his ability to extend his belly in imitation of a thick girthed neighbor named Murphy. Mr. Murphy also went barefoot most of the time even though he was a grown man.

'Buzzard' was the worst thing that Loyd's brothers could think of on a day in the field. In a moment of gold-bricking, being disguised as a work break, they entertained themselves by calling each other repulsive names. For no reason his name stuck through the rest of his life. Mercifully, this was shortened to the more socially acceptable 'Buzz' in adulthood.

Herchel was alternately called 'Bape' or 'Tuffy or Toughy'. The latter simply because he was as tough as boot leather. He had to be. He was slightly shorter in stature compared to his older brothers and early in life his younger brother, R.C. caught up to him. He neutralized this by being strong and agile with the courage of a bulldog. He had few challengers. While adolescents working in the field, Loyd bragged that he could whip Herchel's butt with one hand tied behind him, on his knees. A big mistake because Jack forced Loyd to follow up on his boast. Loyd told his son David about this incident and David passed it on to me while I was researching for this chapter. Herchel, being mobile and much quicker than the impaired Loyd, put an end to Loyd's braggadocio with dispatch. 'Bape' was derived from another childhood tag, 'Ape'. Herchel was very adept at walking on his hands and doing more chin-ups than any of us. R.C. got angry with Herchel for some hazing incident. His quick anger brought a quick response. In an impulsive burst of reckless abandon, R.C. intended to direct a virulent epithet at Herchel. As he made the first sound, an instant rush of prudence and self-preservation interceded and he changed the word so that the sound blended with his nickname 'Ape' to produce 'Bape' and this one replaced 'Ape' forever.

R.C. also had two nicknames. One was 'Smutter'. The history of this is somewhat clouded but as I recall, it had to do with the name of one of his grade school teachers who gave him a hard time for his lack of academic interests. "Jew' was the other favorite, less used, nickname. I don't know precisely the origin of this racist nickname, but I would guess it had something to do with facial features.

Yes, we were inveterate racist. As children we assumed the values and prejudices of our society, this being the pre-politically correct South. Seemingly, man needs someone we view as lower in the pecking order and though unjustified, we assign those roles to man for the color of his skin. Although deep-seated, our racism was of a passive nature. We wouldn't actively do harm to anyone without reason and we didn't hate people for the color of their skin or nationality. We thought members of the KKK were idiots. We simply assumed the inferiority of others not like us. Of course, this attitude can be as debilitating as outright hatred. The best lesson I received on this topic was taught in an indirect way by Daddy. I was accompanying him on a trip to Mineral Wells. On the way home he picked up a neighbor, an old Negro gentleman who farmed a place near ours. Daddy talked to him with respect, without condescension or fear. From that day I understood that I was not to take some statements of adults as gospel. However, racism continues as a theme in the nickname of Roy.

Roy had darker skin and hair than the other brothers. The skin became much darker in the summer when all of us shucked our shirts. He was a good-looking child and as an adult he was a very handsome man. He acquired a nickname that is one of the most reviled, racist and controversial names in our culture. Even he reluctantly accepted it and signed his letters 'Nig' his entire life. Of course, his wife would have none of this and I had a difficult time changing to the use of Roy. My parents and for the most part, June and I called him Junior for obvious reasons. Neither of us relished the consequences. His older brothers would sometimes refer to him as 'Darky'.

Fortunately for me in this discussion, I was the prototype WASP with white hair and light freckled skin that glowed bright red with anger or embarrassment. These characteristics produced my first obvious nicknames, 'Cotton' and 'Whitie'. These names were short lived. The permanent 'Gopher' was the outgrowth of a child's love for playing in the sand. As a family we harvested pecans for others 'on the halves'. We were working on Uncle Arlander Wallace's creek valley farm when I was about five, too young to be of much value in the work at hand. These sandy fields were dotted with dozens of gopher mounds, providing excellent sand piles for June and me to use for play. I got the name. June got the hugs. How could anyone name a three-year old duplicate of Shirley Temple 'Gopher'?

June was the pet, the long-awaited daughter to complement this brigade of ruffians. She, with long, blond curls, was a precious toddler. She was loved and protected by the family, even by me on occasion. June never had a permanently established nickname. Jack would refer to her as June Bug in teasing. Roy and I would sometimes call her 'Gert' in reference to Mamma's mentally retarded cousin, but that one saw a quick death. Mamma saw to that. We were insensitive to a lot of things. Daddy was called Icky, clandestinely, by the older boys because of his structural resemblance to the fictional character, Ichabod Crane.

The Boys continue to put in their time and look forward to getting out in November. Jack wrote on May 11,

"Curtis is going to Ft. Brown Mon. morn at six o'clock to coach our draftees on rifle firing. Sgt. Ranin wanted all three of us to go, but Lt. Hornsby wouldn't let Buzz go because he is his orderly & only one corporal was allowed & I hadn't had enough experience in drilling & handling men to qualify there. I've just been corporal since last Monday. I really think I'm smart now. I'll get $54 next payday. It took me 5 mos. & 13 days to make it. I guess I was lucky. Some of these guys in the M.G. platoon had four & five years on me, but Sam Whatley [First Sergeant] was the one that helped me out. Well, I'm gonna try to make them a good one but it's not easy with a bunch of yard birds like some of these guys are."

Again, this letter reflects Jack's level of maturity eight days before his twenty-fourth birthday and his continuing concern about money. He was always watchful of his brothers, especially Loyd who was more careless and carefree. He entered this paragraph in his letter of May 18.

"Buzzard is on main guard today. He was guarding two prisoners to and from the mess hall. He was carrying a shotgun on his shoulder and it went off. Some of the guys that were there close said he didn't even look around or stop walking. He was really keeping his eyes on those prisoners." This was real poise for an eighteen year old.

Their squadron, the Second, left Ft. Ringgold in late May and made stops at Ft. Clark, Ft. McIntosh and Ft. D.A. Russell before arriving at Ft. Bliss. Jack wrote May 29, his birthday, from Ft. D.A. Russell near Marfa.

"We passed through most of the Davis Mts. yesterday & it was really pretty around Alpine & across the Pecos Canyon down by the mouth of the river. I threw you out a letter yesterday to a kid. I guess he mailed it. This is really wild bare country. Not a thing but hills & prairies. It's like the Iraan [Texas] country but bigger hills."

Jack was always interested in his friends from home. He wrote to many of them. His best friend was Rowe Howard. He included the following paragraph in his first letter from Ft. Bliss, June 3.

"Just as soon as you hear where [Rowe] went let me know. I guess he'll go to Ft. Sill. Click said Lillie & Everett went to see Fabe the other day & that Fabe and Alton were both getting fat. Well I guess they will get along. Ole Kickie King is in the 8th engineers out here now. We saw preacher [Keith Gaylor] but we haven't seen Glen & Loyd Gilbert yet. I think we'll go over there next Sun. They have the tents we had before & we are in a new place. It's rocky as heck. I haven't had time to turn around since we've been here."

I suppose this busy schedule had something to do with a comment that Curtis wrote at the same time. "We had a big review the other day. They said that it was the largest review ever held in the U.S. and that we were the best troop in the review. About 50 troops took part in it, so I guess that makes us pretty good."

Jack was well disciplined and tough, but his longing for home was expressed from time to time in his letters such as this excerpt from his June 9, 1941 letter.

"Boy I'd like to be there & go on some of those fishing sprees, but I'll just have to wait awhile. It's just about five more mos. till we can tell if we are gonna get out or not. They will have to say you're dismissed & hand me a discharge before I believe it tho."

Daddy

Mama

The Family, 1931

CHAPTER 4

FARMERS AND SOLDIERS

AS THE BOYS MOVED with their regiment to the forts along the Rio Grande and settled in to serve the remainder of their year, Germany continued its aggression into the Scandinavian countries and into Greece. On May 20, 1941 Germany used parachute troops for the first time on a large scale in flushing the British from Crete. Although militarily successful, Hitler lost interest in airborne action because of the heavy casualties incurred by German forces.

Rudolph Hess made his insane flight to Ireland in a mysterious attempt to effect the direction of the war. Although his exact goal remains a mystery, some speculate that the trip was an attempt to influence the English to make peace with Germany. This could ease the burden on Germany in the west as it prepared to invade Russia and make a hero of Hess.

Hitler, the impulsive strategist, decided to abandon his plans to invade across the channel after the disastrous air campaign over England in 1940 failed. Ignoring both the advice of history and his own generals, he launched one of the most massive attacks in the history of warfare against the Soviet Union in June 1941. This action was arguably the most disastrous decision he made during the war, perhaps even more so than the declaration of war on the United States later in December. (WWII, American Heritage)

The United States, with nudging from President Roosevelt was beginning to stir even more from its long slumber and realize that in this world, isolation was no longer an option. The Boys journeyed to Fort Bliss for more training for the massive war games to be held in Louisiana and East Texas during August and September 1941.

Defying the odds made worse by drought and The Great Depression, Roy and Martha continued the struggle to maintain their chosen way of life on the farm. Both had siblings who had abandoned the farm to improve their lot by working in the oil fields of West Texas, at brickyards or service stations, anything to assure at least a meager income. Some of them did so well working the oil patches that they could afford new cars. Boy, were they rich! Others within the last year had taken to helping build the army camps springing up all over the country. Mamma and Daddy, even with urging from Grandpa Knight to move to town and get a job, weren't apt to give in.

Not only was the Knight family limited to subsistence farming, they were also cast as tenant farmers. Because we owned our equipment and draft animals we were one slight cut above the sharecroppers who owned nothing but their clothes, furniture and other personal possessions. Tenant farmers were also obligated to share the products of the soil with landowners on a pre-arranged percentage. In the brief period of my awareness of these transactions, it entailed a small cash payment in the fall after harvest, based on the sale of the cash crops such as cotton and the estimated value of the feed crops of corn and maize.

As the changing seasons pull the migratory birds from one part of the hemisphere to another, farmers are pushed to the fields. One can smell the broken soil before the plow hits it. Early in the spring Daddy would float a bank loan sufficient to buy seed and fertilizer for the field crops. We saved seed from last years crop of watermelons, cantaloupes and garden vegetables. The vegetable garden along with the watermelons, cantaloupes and sweet potatoes received the natural organic fertilizer collected from the cow pen and chicken house. In a good year, he would pay off the loan, the landowner and with the remainder, buy winter clothes and food supplies that weren't grown and preserved on the farm.

Horse and/or mule-drawn turning plows tilled the soil. These simple machines consisted of large pointed steel blades, the shares, joined and fashioned with the moldboard to form a curved blade to turn the soil in one direction to the right. Two handles were attached to the back of the plow so that the operator could manipulate to control the direction and depth of the furrow while walking behind in the wake of the plow. The plow had a slide that ran along the ground to help balance and control the plow. The beam, made of heavy, curved steel was attached to the double-tree, a large oak pole with metal rings and hooks at the end and middle used to attach it to the plow and onto which were fastened single-trees at the ends. The singletrees were shorter versions of the doubletree, to which were hooked the trace chains from the harness of the draft animals. Large, thick, leather-covered collars were placed around the animals' necks and fitted against their shoulders. Chains were attached to wood and metal stocks that were fitted against the collars and extended to the single trees. Power was thus transferred from the forward motion of the animal to the point of the plow. In effect, the team actually pushed the plow.

The animals were fitted with bridles. Long, leather straps were attached to rings at the ends of the mouth bits used by the driver to control the animals. Tilling the soil in this manner turned under any remaining plant life and helped aerate the soil.

Daddy assumed the role of plowman until the sons were large enough and strong enough to control the team and handle the plow. This became one of the many rights-of-passage for boys growing to men on the primitive farm. There was little difference between the methods we employed and the practices of the pioneers who settled the North Texas territory in the last half of the nineteenth century.

My first training came from following Daddy as he drove the team and turned the soil. I can still smell the freshly opened soil. There were also great treasures that I might discover such as a rare Indian arrowhead or other artifact plus grubs or unusual stones. There was also the unforgettable pungent odor of the horses' fecal droppings and the sweat of the animals, not quite unpleasant.

When it came time to plant the row crops, the horses were hooked up to the middle buster or, as I recall, the 'lister'. This plow had a curved blade on each side and after making two passes, a seedbed was formed between two furrows. It's height, consistency and true course were dependent upon the skill of the team and plowman. The teams had to learn simple commands, both verbal and physical. 'Gee' meant to turn left and 'haw' was the command to turn right. Although I personally learned to gee-haw, I never learned all about handling a team. By the time I was old enough to be the plowman, Daddy had cut his farming operation and the work force was reduced to him plowing and me hoeing.

Riding planters were used to drop the seeds. This was a machine pulled by a team of draft animals. It straddled the seedbed that had been created. I was fascinated by the synchronization of the wheels and gears with the seeding plates in the seed box to regulate the spacing. A tiny plow would open a small furrow to receive the seeds and the trailing plows, one on each side of the small furrow, would cover the seeds that dropped through a pipe to the soil. In my unsophisticated mind, this machine was an engineering masterpiece of advanced technology. In the smaller garden plots, this work was done with a small plow, the Georgia Stock and by hand with a gooseneck hoe. The seeds were dropped by hand in the furrow. Watermelon and cantaloupe seeds were placed by hand in small individual hills, two or three seeds to the hill.

With luck, rains would come in sufficient quantity and at proper intervals to see a crop through from seedlings in May and June to the fall harvest from September to November. Vegetable gardens were put in earlier and harvested in May and June. We always got enough rain during the spring for the gardens. Long hours of hard work were needed to assure that every drop of water that fell was available for the crops. We cherished the great fragrance that arose from the freshly plowed young crops after a spring thundershower. Grass and weeds became the ever-present enemy and had to be eliminated. The soil must be stirred to break up the capillary action in the soil that brought moisture to the top, causing evaporation, wasting the soil water and stunting the crops.

The riding cultivator was the first line of attack, followed closely with the wielding of the gooseneck hoe. Daddy and sometimes the oldest boy had the 'easy' job of riding the cultivator that straddled the corn and cotton rows. The small winged plows of the cultivator would run the middle between the stalks.

Long levers with handles attached to the plows regulated the height of the plows and the movement of the seat by the operator provided left and right direction. This operation would uproot or smother most of the smallest weeds and grasses. The team had to be muzzled to prevent them from feeding on the succulent corn tops, a wound that ruined the plant.

The 'hard' job was relegated to the rest of us, the 'hoers', to skillfully work closely around the plant stalks, without cutting them down or damaging them, to get the enemy that was left alive by the cultivator. This job was awful until we came upon a patch of Johnson grass. This stuff was a nightmare to us grunts. If nurtured and left to grow at will, this grass could be an excellent hay crop, but it was public enemy number one to row crops. No crop would grow in its midst. The roots ran deep, developing rhizomes in thick networks that choked the life from the cotton and corn plants. Sometimes, for misdeeds around the house, Daddy would sentence us to a work detail to dig out these roots to feed to the hogs. To a pre-adolescent this was a sentence to slave labor. The hours stretched and days were unending. Unknowingly, but with certainty, these conditions produced a maxim for me that I passed on to my three sons: 'adversity builds character'. They still tease me about this. I reached manhood in spite of it and have had challenges much greater than Johnson grass.

At the same time that we were working to produce the staple crops, we were also raising our vegetables and Daddy's much cherished watermelons and cantaloupes. These were fun crops for the family and provided some cash and an opportunity for Daddy to 'get away' and practice one of his most accomplished skills, salesmanship. In the early years he and the boys would load a horse-drawn wagon with these fruits, roastin' ears from the corn crop and various excess garden vegetables and head for the city. Included were the purple hull peas from the field that were planted every third row in the corn to provide nitrogen enrichment for the soil. Mamma would take along excess butter and eggs hoarded during the week.

Mineral Wells, a resort city famous for its 'Crazy Water' was the closest town and possessed the out-sized Baker Hotel and the Crazy Hotel. Daddy would first ply his goods at the hotels where he had cultivated friendships with chefs. Produce not sold to the hotels would be peddled to the grocery stores and then from door to door to selected residents. Rarely did we need to bring anything home. If we did, we ate it or canned it. On the rare occasions when I was the chosen sidekick, I learned a lot about salesmanship and integrity. Daddy always guaranteed his produce and stuck to it.

Daddy's friendship with the chef at the Baker Hotel connected him with Dallas businessmen who provided him with yet another source of income. They hired him as a guide for duck hunting on Lake Mineral Wells. He received a new Remington twelve-gauge shotgun from them as a bonus for his efforts in building blinds and paddling the boats around the lake. The gun was passed on to me near the end of his life. I have passed it on to Scott, my middle son.

These peddling trips occurred on Saturday for the most part and if Mamma and June went along, the oldest boy was left in charge of the gang at home. Daddy recognized the need for his 'work crew' of sons to have some leisure time. These loosely supervised days led to adventures that were always intriguing and sometimes dangerous. Our activities usually centered on fishing and hunting and swimming. We also practiced war games with rubber guns or corncobs. The 'rubber guns' were actually constructed of wood and a clothespin rigged to release a stretched rubber band cut from an old inner tube. Both of these activities could be injurious, but none of us ever lost an eye in the ferocious battles that took place. We weren't environmentally sensitive and would kill anything that moved by any means available; .22 rifles, slingshots, small rocks, sticks or our pet dogs, Ted and Bruno. There were two types of slingshots; those constructed from a forked stick and rubber bands that folks in the South gave an ethnic nickname and there was the biblical type used by David to kill the giant. It was made from two long pieces of cord and a small leather patch tied to the ends. It was slung in an arc around the head with one string attached to the wrist and the other released to launch the stone. Much practice was needed for accuracy with either weapon, but I never mastered the skill to hit anything with consistency with the latter. I couldn't emulate young David, perhaps because we had no giant enemies for motivation.

In August, when the crops were sufficiently mature to survive and produce, they were 'laid by'. We quit hoeing. The heat and lack of moisture would place the garden in decline and we would have a respite until the fall harvest of cotton and corn. The heat would become so intense that we would move our beds outside under the shade trees to escape the captured heat in the house and to catch any breeze that came our way. Daddy would try to get a nap at noon, but the racket created by mocking birds and scissortails made this a futile activity on most days. I learned some watered-down cuss words during his rampages. The birds always averted the various missiles he launched in their direction. I have yet to learn the meaning of one of his favorite epithets, 'Dad shiest!' He had friends and relatives who served in France during World War I and he may have picked up some G.I. German phrases. However, at the time, I just accepted it as something he was allowed to say, but I had better not use it. During these August days we would engage in the sporting phase of food gathering along with swimming, making ice cream and picking stickers from the leather-like soles of our bare feet.

Daddy had the eyesight of a hawk. One of his favorite pastimes was hunting for bee trees with Wade Howard, his best friend. I received my middle name from his son, Rowe. Many times I listened to them argue and discuss politics as I sat between Daddy's legs on the front porch. I usually preferred this to playing with the other kids. I could play with kids anytime.

Daddy and Wade would find the bees on flowers or around water holes and watch them as they flew away to 'course' the direction of the bees true flight to their natural hives in large hollow trees. They would dust one's wings with wheat flour and time its return to estimate the distance to the hive. With this intelligence they would walk up to a mile toward the tree to skylight these tiny insects up to fifty feet in the air as the sun glistened from their wings. They came from all directions, swarming into the hollow in rapid succession. The cry, "Here it is!" became a familiar call that a bee tree had been found. I never developed the skill to see these little creatures because I kept looking for a much larger form.

The healthiest hives, determined by the level of activity, would be robbed by cutting down the trees with a crosscut saw and by cutting out large chunks of the tree trunk to expose the honeycomb. Using the usual methods and attire of bee-keepers, the honey comb would be cut away and carried home in wash tubs to be transferred to Mason jars. Trees were robbed in August and September and sufficient honey would be left for the recovery of the hive and winter survival. Honey along with preserves and jellies of peaches, blackberries, grapes or plums spread on one of Mamma's large, hot, butter-laden buttermilk biscuits provided an exquisite culinary experience.

My life was filled with lessons, not always pleasant. One of these came at a time later when I was old enough to help Daddy cut a bee tree and I was the only help available. Late into the 1940's, after all 'The Boys' but me had left home, Daddy and I continued this sport. At one of our bee-tree cuttings I learned one of my most valuable lessons. It was a typical Texas August, hot as hell.

I thought I had reached a reasonable level of studness and was ready to claim another right-of-passage. Daddy could make a chew of Beech Nut chewing tobacco look exceptionally delicious and I knew that only a real man could handle the stuff. I had become somewhat of a risk taker. With only minor trepidation I ventured, "Daddy, how about a chew of that Beech Nut?" He looked at me and without a word, reached into his overalls bib and extracted the red and white sunburst designed packet of this loose leaf, sweet tasting narcotic. With only a hint of mischief he pulled out a generous portion and handed it to me. Of course, I thought I had to stuff the whole batch in my mouth. As I labored to get this mass of treated vegetation into a controllable cud, I was unconscious of the trickles of juice flowing into my shocked gastro-intestinal system. As I continued to chew and my salivary glands continued to work overtime at full production to meet the demands of this foreign intruder, more and more of this repugnant juice was finding its way to the stomach. My whole system entered a protest mode.

Suddenly, my whole body began to react. I lost my balance. My stomach was in loud and convulsive reaction. I could do nothing but lie down and then the whole valley entered the protest. It began to spin faster and faster. A violent eruption from my stomach sent this vile mixture spewing forth. After my stomach was emptied, I lay back but the world kept spinning. I had never experienced a sickness to closely match this. Death would have been a welcome option. One of my best learning experiences came, not from a lecture, but from Daddy who simply complied with a simple request and never said a word, even in sympathy.

Daddy was a benevolent autocrat. He never made us go from pre-dawn to post-sundown as some farmers did, due in part to his large pool of 'free' labor. He was also a wise and reasonable man. We worked a good day, took off most Saturdays for play or going to town and every Sunday to go to church and rest. Of course, chores had to be done every day. The animals had to be fed and the cows had to be milked and the eggs gathered without fail.

If Mamma needed help around the house, the labor force was divided. If she needed more help during the canning season, all of us got a chance to peel peaches, shell peas, cut corn from the cob, shell and snap beans and wash and sterilize the jars and lids. She operated the pressure cooker that sterilized the contents of the jars and assured a vacuum to seal the lids. She pickled cucumbers in various ways: sour, bread and butter, dill and sweet. The firmer varieties of peaches were pickled without removing the pits and took on a beautiful dark pink color in the jar. I can't forget the rich red of the pickled beets. Food freezing would come later with the advent of rural electrification.

To help preserve food short-term, we used an icebox with home delivered ice or if we lacked the money for ice, Daddy would fashion a cooler in the window. It was made from a large metal pan filled with water and tow sacks hanging in the water from a wooden frame. The water would soak the sacks and the summer breezes would evaporate the water, cooling the interior and keeping the food from getting too warm. It was never chilled. This contraption was used for milk and butter almost exclusively. We ate a lot of clabber. We also had the convenience of farm-to-farm peddling of groceries from a truck and the occasional visit of a Watkins Products salesman.

All of our cooking was done on a wood burning range made of heavy cast iron. On the range was a compartment for the wood to burn. Several lids were located over the fire. These lids could be removed with a metal tool. Skillets and pans could then be exposed directly to more intense heat. After cooking, the lids would be replaced and the food kept warm by the lids heated by the dying coals. The oven was located beside the firebox and the temperature for baking was regulated by the amount of fuel in the firebox and the damper in the stovepipe. The correct temperature was determined in a most unscientific way, by placing a hand into the oven to feel when it was about right. Three hundred fifty degrees was toasty warm. Mamma was fairly proficient at this, but I can still eat charred biscuits, cookies and piecrusts because of the childhood experience of developing a taste for carbon produced by Mamma's miscalculations or distractions. We simply didn't discard burned food unless it was unrecognizable.

After we started school in September, we began the harvest by gathering the corn in the afternoons and on Saturdays to beat the fall rains. Being the youngest, my first experience at harvest time was driving the wagon team. This was initially very difficult for a seven-year-old who was victimized by the disrespect of the team who ignored my timid commands. The hostility of my brothers, (feigned or not I could not chance) on the ground added to my misery. Their anger was kindled by my failure to stop and start the wagon moving at the precise time they deemed most advantageous for their efficiency of labor. I had to be very alert so that those on the ground wouldn't get too far behind. Too much of a transgression would result in an ear of corn being hurled my way. Looking back, I don't believe they intended to hit me because they had the ability. I could have been badly injured and that would have brought the wrath of Daddy upon them. But at the time I was convinced otherwise. I was never hit, but I was always tense.

After I was old enough, at about eight, my job was to pick the 'down row' that was created by the wagon breaking over the cornstalks it straddled. The older boys would pick two rows on each side of the wagon. The team would be left unattended and would be controlled by voice command of the brother with seniority and delegated authority. Was my first year for training or just a way to get me out of the house? The down row created difficult, stooping work even if I was built closest to the ground. I had to keep up without help or any control over the team. I no doubt started the development of my throwing skills when I got behind. Adversity builds character and an arm for baseball. The brothers' skill at renaming the horses in a most profane way was also my most intense indoctrination into the art of cussing and the 'hierarchy of cussin' that I will discuss later.

These skills were handed down from the oldest to the next in line. Curtis told me a story about R.C. learning from him. After working with Curtis, driving the team and listening to his diatribe, R.C. decided to vent his anger at one of the more cantankerous horses, emulating Curtis. He cut loose with "You sonala bitchin' ol' black hole." The line between the discipline of draft animals and language development at times became entwined.

'Pulling boles' was our approach to gathering the cotton crop. Cotton gins had improved so that it was no longer necessary to pick the fiber pods from the bole, a much more tedious and time-consuming process. We pulled a long cotton sack made from heavy cotton ducking by a strap over the shoulder. The boles were picked and shuttled to the sack opening. With sufficient skill and strength and a bountiful crop, a bole puller could place an elbow in the sack opening and keep a continuous roll of cotton going from the quickly moving fingers to the sack. This became much easier after the first frost when the stems became more brittle and the leaves fell away. We received our first training in this job by following Mamma or Daddy, picking a few boles and placing them in their sacks.

When we were gathering our own cotton crop, we weighed the cotton to get an estimate of the weight to avoid possible errors (cheating) at the gin scales. After our crop was in, Daddy would hire us out to other farmers less fortunate in the production of sons. We were paid by the hundred-pound. I don't recall drawing any wages for this work, nor do I remember feeling the need to complain. We just innately knew that this work was necessary to provide all of us with shoes and winter clothing along with other necessities.

Shortly after the fall harvest of crops, we began a vigil for the first chilling cold snap. We could then safely have the annual 'hog killing'. Slaughtering hogs was never a pleasant experience, but to me at ages four to seven it was traumatic. The hogs' experience with humans had always been a rewarding one. A steady supply of food with little effort on their part would bring them to the trough without coaxing. To see them trustingly look up at the rifle toting person, expecting a meal of slop and corn or wheat meal and then to see them shot between the eyes and hear their anguished cries shrunk my soul. This act of murder was followed by sticking the hog in the carotid artery with a large butcher knife. This was done to bleed the hog. The sweet smell and the sight of the warm blood pumping onto the ground in a mist of vapor and hearing their final gurgling protests drove me away.

Hunger could dim the memory, because I never had a problem eating the bacon, ham and sausage that this labor produced. I suppose a mild form of brainwashing occurred over the ensuing years as I was gradually acclimated to the ritual. Killing a hog became another right-of-passage as I took my turn in early adolescence pulling the trigger and wielding the sticking knife. The first time was difficult, but then it was just another thing that adults had to do to survive in the Darwinian world. I had to become a man someday, didn't I?

Before the killing and bleeding of the hogs, we had built a large fire under our cast iron kettle that had been filled with water. Nearby, a hole was dug and a fifty-five gallon metal barrel was placed at a 45 degree angle. The hogs would be dragged to the site of the kettle and drum on a horse drawn sled. The barrel would be filled with scalding water and the two strongest participants would wrestle the hog into the drum of water and turn it over and over until the proper scald had been achieved and then the ends would be reversed to a get a whole-body scald. The animal would be pulled again onto the sled to be scraped clean of hair and dirt with butcher knives that had been honed to sharp perfection by Daddy.

The hog would then be strung up by block and tackle to a nearby tree limb with its hooves hooked to the ends of a singletree, to be gutted with the entrails captured in a large washtub. The edible organs would be removed for immediate consumption along with the spare ribs and pork chops. The gastrointestinal organs would be hauled to the back of the woods as a peace offering to the dominant carnivores and buzzards in the area. Some farmers made use of the stomach and intestines. But Mamma sewed sausage sacks from cotton cloth, usually flour sacks that were collected for this use and for clothing. The hog's head would be made into souse meat loaf. For the most part we cured the hams, shoulders and bacon with a sugar and spice mixture and hung them in the smoke house. We didn't use smoke, but wherever we cured the meat became the 'smoke house', a carry-over from earlier times when smoke was actually used. We would run all the other edible parts through the sausage grinder and mix in the spices. This concoction would be squeezed into the tubular sacks and hung in the 'smoke house'. Hopefully the cold would keep the flies away and the sugar mixture would shield the meat from a variety of fungi until blood and other moisture drained to the point that the microbes could not sustain life. This supply of pork would last well into the spring and was our primary source of meat.

The belly fat would be cut into small squares and placed into the kettle for rendering for lard. When the molten fat was dipped out into five- gallon cans and the rendering process was complete, the cracklings would be saved. Small amounts of the cracklings would be put back for cornbread. Most of the remainder would be mixed with lye and water to make soap that Mamma used for washing clothes.

Washday was another dreaded event. It meant that those of us who were old enough had to haul water to the same cast iron kettle to be heated to boiling. Mamma would shave the lye soap into the boiling water and toss in the white parcels for the first wash, then the light colored fabrics and finally the dirtier work clothes. My job was to keep the fire blazing hot by adding wood as the fire diminished and to poke and stir the boiling clothes as best I could with a long stick. The clothes would be lifted out of the boiling water with the stick and transferred to the first tub of rinse water. Mamma would look for soiled spots that hadn't been removed by the boiling and stirring and work on those spots with the rub board. She would then wring the clothes with my assistance and drop them in the second rinse tub. When June was old enough, she got to help rinse at this stage. The last rinse tub contained water with bluing added. This chemical helped brighten the whites and lightly printed items. By the time the work clothes made their way to last rinse, the bluing had largely lost its effect.

Mamma hung the shirts, blouses and Sunday britches on the clothesline and if more room was needed, the work clothes and older fabrics would be draped across the nearest barbed wire fence. We had to be very careful when removing the dried clothes from the fence to keep from snagging the material.

June and I would help Mamma sprinkle the clothes to be ironed. She used steel irons heated on the cook stove. I would never develop this skill. By the time I was old enough to reach the ironing board and acquired the manual dexterity, we had moved and acquired electricity. We only had to move a dial to get the correct heat for the fabric being pressed. I can't lend any expertise in what had to be the fine art of gauging the desired heat on the old flat irons, but it entailed the use of spit on a finger touched to the iron surface.

Our family lived on several farms over a period of thirty years. In the winter we had difficulty keeping the old clapboard houses warm. The structures were simple and flimsy. There was enough frame to nail on one by twelve pine boards vertically with one by fours nailed over the seams. No caulking was used and no paint applied. A painted house represented to me another social stratum beyond us. As the wood dried the seams enlarged. Nails had to be driven back in and others added. The inside walls were covered with gauze tacked to the boards on which was glued the wallpaper. The paper served two purposes. The colorful, patterned paper provided a somewhat questionable aesthetic touch. Layer upon layer of the paper added by various tenants also had some effect as insulation against the intruding winter winds. Aside from the howling of the dogs, the mournful whistling of this cold wind through cracks in the house accompanied by the bellowing of the wallpaper was the definition of forlorn and lonely.

One of these houses, the Holder home place where Mamma grew up, had a stone fireplace that provided heat. The other houses had to be heated with a wood-burning heater. We didn't have a cast iron heater that would retain heat late into the night. Ours was a thin-walled sheet metal design that cooled off rapidly as the fire and the coals dimmed. We kept warm in the late hours by sharing body heat of two or three to the bed, insulated by layers of blankets and handcrafted patchwork quilts and comforters. On the very coldest of nights, Mamma would wrap hot bricks and place them at our feet.

Another unwelcome source of warmth was the ever-present chance that one of us younger ones, usually me, would pee the bed. Of course, this led to the misery of all occupants as the warmth faded and the icy reality would join the pungent odor to create indescribable discomfort, and verbal recriminations with equally vehement denials. I always lost those arguments. The littlest Knight didn't have a chance for justice. The final solution was to jump out of bed, run to the fireplace or heater where Daddy had built a fresh fire and strip off the wet woolen long handles. Most of the time I slept so soundly that I didn't know my condition until Mamma or Daddy woke us to have breakfast and get ready for school. The oldest boys had to help milk and feed the stock before we departed to catch the bus for school.

The Boys were in the midst of their advanced training at Ft. Bliss, preparing for the coming maneuvers in Louisiana and counting the days until they could get out of the army and come home to resume their lives. On June 19, 1941, Jack wrote,

"We made a big review Wed. & the General said the 124th was the best riders in the 1st Division. Glen & Loyd were in it. The papers said there were 15,000 men in it but General Pyron said there was about 12,000. The whole drill field was full & it's a big field.'

"Well we have 4 mos. & 29 days or close to it. We will be busy from now on & I like it because the time will pass off faster. I don't think we will get into a war very soon do you all? We can't ever tell tho. I believe if we do that we will go back to the border.'

"We got in six Yankees today. I'll bet they hate this place. Most of them are from Brooklyn, N.Y. They talk like Wops or something.'...

The troops kept hoping that their year would be all they would serve. With the isolationist mood in the country and the public statements of government officials, hope wasn't unrealistic. But I believe that deep in their minds and guts that they knew that we couldn't remain neutral.

By late June, after preparations and parades, the units took to the field. Jack wrote from a 'Wind Swept Desert' in late June,

"We haven't seen Buzz since Thursday morning about 4:00. He is out here somewhere with the 5th Cav. We have been fighting them but I never saw him. We are going over on their side sometime this week & fight one of the other Brigades. We are really having the fun. We went from 2:30 one morning till 3:00 the next morning without a dang bite to eat. The Capt. like to have died. He is the biggest sissie in the troop. We just had a canteen full of water (1 qt.) but it rained & we all drank out of the ditches. I don't think the Capt. would drink that rain water was the reason he took it so hard. I really like it. The rougher it gets the better I like it. If I can make these officers suffer its fun to me."

He was telling Mamma that he was doing well and not to worry and at the same time telling Daddy that he raised sons who were tough enough to take anything the army could dish out. There was also an implicit message that he was still a workingman with contempt for the bosses.

I recall the pride that Daddy tried to instill in us at the expense of the rich and privileged. He could make me sympathetic toward the rich city kids who didn't have the privileges of the rural life like we had. Even today, there is enough truth in those concepts to give it validity.

In his letter of June 29, Jack gave a brief account of one of their skirmishes that included the following statement.

"The First Cavalry Division used a batallion of tanks on us, 13 of them they ran right through our horses where we had them back of the battle line & scattered them all over the place. We had lost lots of sleep."

This action was foreshadowing for coming events and an indication of the difficulty that the use of horses in modern warfare was facing. They had serious business at Ft. Bliss trying to prove that horse cavalry could remain a viable part of military tactics. There was some leisure time left to the troopers as related in Jack's July 6 letter.

"Buzzard said that the water melons were real late this year. We go to town every once in awhile & eat watermelon. You can get a good big slice for a dime & its better than eating ice cream. We went Friday night & Jack Stockstill went with us. He & Buzz acted a fool in the show for 2 solid hours & then got out & eat water mellon till they couldn't. They call all the girls that work in the cafes,

Myrtle & Girt & Emmer. They made one old gal mad but couldn't do anything about it. I don't think I'll go with them anymore. They are too crazy." We were doing well to have ripe watermelons by July 4.

In Jack's letter of July 14,

"Curtis and I went to Carlsbad Caverns yesterday with Click, Noble Stockstill, J.C. Byrom & Lowell Holt. We got Paul Crawford's car & it cost us $2.85 each for eats, gas, drinks, candy & a watermelon. Boy there was really a crowd there. I got a folding post card. I'll send it home. I've got some I got in Juarez, too."

The Boys were enjoying this first meager experience of affluence. Before this they rarely had $2.85 at all, much less being able to spend it on pleasure. Jack had saved some money before joining the cavalry, but he wouldn't allow himself to be self-indulgent. Loyd's intelligence and charisma had a way of landing him in some interesting & less taxing assignments. From Jack's July 14 letter:

"We made our ride fine last week. It is better to be out in the field. We don't have to clean our equipment & we get off almost every evening. We are going out again Thursday & stay till next Thursday or Friday. Buzz has already gone. He & Lt. Hornsby left yesterday. They went out with a bunch of other officers that are going to umpire the manouver to set up command posts & look the ground over. That sorry Buzz won't do a thing but sit up on a hill somewhere with Lt. & watch the fights & hold his & Hornsby's horse."

Jack added at the end of his letter, "It looks like we're going to be stuck in here for some time now. Today's paper said the President was in favor of it. Well we'll have to make the best of it I guess."

Our troopers were aware of the progress of the war in Europe. In late July when the Soviet resistance to the invading German armies was beginning to stiffen, Jack commented at the end of his July 25 letter, "I forgot to mention it but it seems that our friend Mr. Hitler has kinda bumped a stump doesn't it. I wish he'd drop dead this minute."

Jack's description of their last day of maneuvers at Ft. Bliss contains some more clues to the questions the military had about the use of horses in present day war.

"Well this is the last day of the Ft. Bliss maneuvers. We came in from Desert this morning. We are just getting back to camp & resting at the same time. It is only about 15 miles back where we stayed last night. We got here about 10:00 this morning & have been resting all day. I guess Curtis told you about it but if he didn't, we really had some long rides. We rode about 50 miles one day. We have lost about 25 or 30 men & about the same number horses. All the horses were either crippled or wind broken or had saddle sores. A lot of the men went back because their horses went out on them. My horse went lame & I rode a truck on & caught up after about 10 or 15 miles. Then when Curtis got his foot stepped on I took his horse. His foot is o.k. now. He was on stable detail today (which is Sunday). I've been so busy the last two days I didn't have time to write. Well we got back o.k. Lots of shooting & riding and staying up late & getting up early & water calls & acting a fool in general but no serious misshaps. Some of the boys lost their guts & fell out to rest or do anything but go on. They would try to cripple their horses or run them out so they could ride the trucks. That Buzzard was with the 5th Cav. & they made a surprise attack on us one night & wiped us out. He started riding that day at 7:30 & rode 73 miles by 3:30 the next day. He was swapping from his horse to Lt. Hornsby's & leading the one he wasn't riding. That was partly after the last battle was over. They came from a way out 38 miles in the Hueco Mts. & attacked us at Desert & then went on North to Ora Grande that night & then he came all the way back to camp the next day. Lt. Hornsby rode an umpires car back to the Post'...

"Lt. Hornsby gave Buz a good eating out for riding their horses so hard but he was bragging to me about it. He thinks Buz is pleanty tough but he just ought to see what is coming up at home. That Bape & Smut & N[.....] & Whitie will make us look like a bunch of camp fire girls."

CHAPTER 5

A GREAT GENERATION, EXTENDED

THE BOYS, NOT BY any design, found themselves in a military unit that differed little from ones used a century before and, except for the weapons, much like those of ancient times. Their physical prowess, eyesight and coordination had been passed to them by their ancestors. The most recent was their great-grandfathers on both sides of the family. Jack wrote the following accounts in a paper for his college English class (c. 1937). He first described the migration of the Holder family to the United States, which occurred long before the Revolutionary War. As the generations passed, the Holders had moved to eastern Tennessee and further to western Tennessee and the tale of Isaac Holder began. Jack was fortunate to have a great aunt, Jennie Baker, living in Weatherford with a rich store of oral history to pass on.

"The first born was a son which they named Isaac. I do not know wheather his father was named Isaac or not, but he probably was. It seems that in those days it was a popular custom to name one of the boys of the family for his father. This boy's grandfather had been named Isaac, and I guess that this was his fathers name.'

"Soon after the son was born his father was called away to fight for the independence of the colonies. He never returned. His wife was so broken hearted that she lived only a few months. Isaac was only four years old when his mother died. He was taken in by his mother's family and reared to manhood by an expert hand. He married when he was about twenty two or three years old, and moved away from his family into Southern Tennessee, where he lived when the War of 1812 broke out. He joined the ranks of the United States army and was in Jackson's army in the Battle of New Orlines.'

"On his return home he found that his wife with the whole neighborhood had been killed by the Creek Indians. They were not the only ones to be killed, but there had been hundreds of families murdered in this outburst of hostility.'

"As soon as Andrew Jackson heard about this he gathered together an army of the fathers, sons, husbands and brothers of those who had been killed and went after the Indians.'

"The almost insane army of men found and slaughtered several hundred of the Creeks and ran the others into Florida, where they captured the chief and hanged him. Florida was Spanish territory at this time and Andrew Jackson came near drawing his nation into another war by going into this territory after the Indians. Spain, however was not anxious to get mixed up in any war with these Indian killers for fear they might get the same thing the Indians had.'

"Isaac Holder returned to his farm and started, once again, to work. He lived here several years before he married again.'

"He lived to be a very old man and died the father of six children, three boys and three girls. One of these boys he named Isaac, keeping the old name that his grandfather had brought with him across the Atlantic.'

"Isaac the fourth was very much like his great grandfather. He was over six feet tall, very dark skinned, black eyed and black haired. He grew up with the family but was very different from the others. He was carefree and restless. He would pick fights just for the fun of fighting and it has been told that he never was whipped. Of corse he never ran on to the right man. Once while on one of his sprees he got off a good ways from home. While he was on his way home he came to a house where there lived a girl who was wanting to get married. She already had the man but preachers were few and far between in those days. The fun loving Isaac told them he was a preacher. They agreed to let him perform the ceremony. He ate the wedding feast, got two dollars for his work and went happily on his way.'...

"When the Civil War broke out Isaac Holder was a middle aged man. This, however did not keep him from joining the southern Army. He stayed gone four years, and never got so much as a slight wound during the whole of his four years. He came back to his family where he stayed until 1870 when his old restless spirit showed again. He loaded what he could on two wagons sold the remainder of his property and started west for Texas with some of his neighbors.'...

"The name Knight is a very old one. As much as I'd like to I can't connect myself with some gallant gentlemen of the Medieval times; instead the very first of my ancestors that I know anything about was born about the year 1835. He was my great grandfather. His name was Abner Knight. He married when he was about twenty years old and had a son three years old when the Civil War broke out. He lived in Southern Alabama and; therefore joined the southern Army. He came home one more time about 1862, but was off to the war again in a few days, never to return. A friend of his, later, told my great grandmother that he had been captured and taken prisoner to a northern fort on the Mississippi. He and two other prisoners took advantage of a guard and took his gun. They slipped out of the cell and got into some underbrush nearby and escaped. They built a raft of logs and by this crude method of travel started off down the Mississippi River. They were never heard of nor seen again.'

"The son that this man left at home bore the name of Robert. He was the only child and likewise the only one of the name Knight.'

"When Robert was about twelve years old his mother married a German. The German was a shoe cobbler and my grandfather spent the remainder of his boyhood in Burmingham, Alabama. When he was about twenty his stepfather and mother moved to Texas."...

Jack had great pride in his ancestry that came close to matching his pride in his immediate family. In one of his letters from Fort Bliss, Jack had teased us boys at home about how tough we were going to be. His light-hearted comment would fit Roy, who completed high school in 1947 at the age of sixteen. He had decided before his senior year in high school to become a scholar and made exemplary grades. He tried for a scholarship at Texas A & M, but his whole high school record had to be evaluated. He failed to receive the scholarship. He then chose to work until his seventeenth birthday. With reluctant permission from Daddy and Mamma he joined the Air Force in February 1948. He served in administrative work and reached the grade of staff sergeant. In 1953, after deciding upon a career in the Air Force, he completed officer candidate school and resumed working in administration. In 1958 he determined that he must attend flight school if he was to advance as an officer.

Roy completed his flight training and received his wings at Laughlin Air Force Base in Laredo, Texas in 1958. He was named the outstanding military student in his class. He was assigned to fly jet fighters. Roy had married Patricia Henderson in 1953 and they had two children by this time: Roy III (Chip) and Gayann. Bryan was born two years later while Roy was on a tour of duty in France. He became a flight instructor and served in that capacity until 1966.

The Viet Nam War was heating up with the ill-advised escalation ordered by President L. B. Johnson. I asked Roy to resign to avoid an assignment in the 'sausage grinder' of Southeast Asia. He quickly let me know that he was a professional who had signed on for the work of a warrior and he wasn't running when war was on.

His time came in the fall of 1966. He received orders for assignment to a school to check out in the World War II vintage A1E Skyraider and for training in search and rescue tactics. In January, 1967 he left for survivor training in Utah and was then assigned to the 602nd Fighter Squadron (Commando) at Udorn Royal Thai Air Force Base. He began flying missions on March 11, 1967. The primary mission of his squadron was search and rescue of pilots shot down in enemy territory. The unit filled its spare time on bombing missions and cover for ground combat.

His wife Patricia and their children, Chip, Gay and Bryan were living in a house Roy had built on the family farm. My sister June, lived about a hundred yards north and my parents lived about a quarter mile south. My family and I had moved about 150 miles to the east to take my first school superintendent job at Saltillo, Texas. I had enrolled in the school administration doctoral program at East Texas State University. All of us were trying to carry on our lives, hoping he would return in the fall. Loyd was retiring and preparing to move to the old home on the family farm.

Roy Jr.
At Udorn, Thailand

Starting a mission

One week after joining his squadron, Roy had an experience that would demonstrate his mettle and set the tone for his brief tour. For his action he was awarded The Air Medal. His citation read:

"Major Roy A. Knight, Jr., distinguished himself by meritorious achievement while participating in aerial flight as pilot of an A-1E Skyraider at Udorn Royal Air Base, Thailand, on 18 March 1967. On that date, Major Knight had just taken off in his bomb laden A-1E Skyraider when his cockpit filled with dense black smoke, obscuring his vision. He concluded that he had an engine fire and that engine failure was imminent. He managed to jettison his fuel tank in the only clear area available and elected to retain his complete ordinance load despite the fact that it meant certain death in the event of a crash landing in the heavily populated area surrounding the base. He was successful in effecting a safe landing and averting a tragedy. The professional skill and airmanship displayed by Major Knight reflect great credit upon himself and the United States Air Force."

His wife, Patricia told me that if he had elected to drop his ordinance at the time required, he would have annihilated a chicken farmer and his flock. The air medal is equivalent to the Army Bronze Star. He was to earn seven Air Medals during his two months of action. The citation that covered five of these medals reads in part:

"...During this period, outstanding airmanship and courage were exhibited in the successful accomplishment of important missions under extremely hazardous conditions including the continuous possibility of hostile ground fire. His highly professional efforts contributed materially to the mission of the United States Air Force in Southeast Asia. The professional ability and outstanding aerial accomplishments of Major Knight reflect great credit upon himself and the United States Air Force."

Roy continually wrote letters to his family and made tape recordings for his wife. On April 3 he wrote a letter to Loyd. He needed to vent and he chose Loyd who was a professional army officer and about to retire as a Lt. Colonel. He knew he would understand. The following is an excerpt from that letter, beginning with concerns with the deteriorating condition of Daddy's health. Daddy had become increasingly erratic in his behavior, especially in the operation of his beloved red and white Ford pickup. We were beginning to see the need for medical advice.

"...Received your letter a couple of days ago along with a couple from Pat that relieved my mind considerably about things at home. I probably got more excited than I should have but I was really talking to myself for a few days around here. I should have rested easy knowing that you, Bill, Pat & the others would see that things went right. At any rate I'm certainly happy that things appear to be picking up. Pat reports that Daddy is much happier now than she has seen him in a long time."...

"I've been pretty busy the last couple of days. Was on alert (SAR) [Search and Rescue] and had quite a bit of activity. Went over to N.V.N. just north of the DMZ to try to get a F-105 jock out who had been shot down. We were just 30 minutes late as I figure it now. He was alive & talking to his wingman on his survival radio until just before I got there with choppers and firepower. He was getting fire (small arms) from both sides of a ridge he was on and after I got there I never heard another word from him. The Navy was there with some A-1's and Jets and were dropping 500 lb. bombs right next to where they thought he was. So if they didn't kill him I'm sure he was captured. I got so involved with trying to find him that I probably over-exposed myself-though that's what you have to do when you're low lead of the rescue forces. At any rate I was probably lucky that I didn't get shot down. I did pick up a couple of 30 cal. slugs. One cut some hydraulic lines and the other knocked out a small navigational receiver. No real problem. I completed the mission and flew a follow up in the same area today. We never did get anything out of him today. Some Jets heard a beeper but I'm convinced that it was some gook trying to lure someone within range of the 37 mm guns that got the 105 & missed me yesterday.'

"I guess it sounds like I'm telling war stories, but I kinda feel like talking to someone about some of the things that go on and I know I can trust you to keep it to yourself. I don't tell Pat anything except that I fly missions-none of the flak and ground fire stuff. I'd rather not trouble her. She has enough to worry about as it is. This is really a shitty war for a lot of people. The army undoubtedly has the worse end of it by far. We have some people in the outfit that have worked with Army units as FAC's [forward air controllers] and they have nothing but high regard for the ground troops.'

"Our own people really have a bad time up north. That whole country is covered with missile sites and we really lose a lot of aircraft. So many of them are in areas that it's suicide to try to rescue and there's just no hope for them. I really wanted to fly a high performance aircraft when I knew I was coming over here but now I feel like I've got as good a job as can be had. Comparatively we're as safe if not more so and we're effective in our strikes as well as having a good 'rep' with the jets for our rescue work. I really can't complain.'

I am sure that the action he described in this letter is what led to his receiving the Silver Star. The citation reads:

"Major Roy A. Knight, Jr., distinguished himself by gallantry in connection with military operations against an opposing armed force as a A-1E Skyraider Pilot in North Vietnam on 2 April 1967. On that date, Major Knight entered an area of extremely heavy and accurate antiaircraft fire in a low-level search for a downed crewman. Working at altitudes of less than two hundred feet in withering hostile fire for more than an hour, he continually exposed himself in an attempt to make visual contact with the pilot. It was only after darkness and the loss of radio contact made any further attempt futile that Major Knight withdrew from the area. His complete disregard for his own safety and outstanding courage in the face of overwhelming odds reflect the highest standards required of an Air Force pilot. By his gallantry and devotion to duty, Major Knight reflected great credit upon himself and the United States Air Force."

Loyd wrote to Roy to keep him abreast of the quickly developing condition of Daddy. His detailed letter of April 10, 1967,

"Dear Nig,

I waited to write again until after we had the final word from the Dr. at Scott & White. Curtis & I went over there Sat. morning & talked to him. I left there and went directly to W'ford then to M'sap for a rather hectic weekend of passing the word to everyone.'

"The Dr. told us that Daddy was in good physical condition for a man of his age. His main trouble is loss of memory. He said he was not "crazy", but far from it, that he is smart enough to realize that his memory is failing and that it is terrifying him and makes him extremely nervous. He said that Daddy was trying to cover up his fear by means of continuous joking and somewhat rambling, and at times, confusing conversations. He said the deterioration of his memory was being caused by improper function of blood arteries in his head and that he expected that it would be rapid and that we could look forward to having to take over more & more of his business for him. He didn't specify any particular time element that someone would have to take care of his business completely but I got the impression that he was thinking in terms of 1 or 2 years from now. He said there was nothing that could be done to stop or help his condition, that it was prevalent in a large percentage of people about his age and older. They gave him, or prescribed, medication that has a calming effect and causes him to take frequent naps and sleep better. I asked the Dr. if he thought Daddy was capable of violence to the extent that he would be dangerous to anybody and he said that in his opinion, "no". I also asked him what effect it would have on him for members of the family to try to influence his mannerisms by "scolding" him. He said it would definitely have an inimical effect and should be strictly avoided. I passed the word to all concerned on this subject. Curtis asked about his driving and the Dr. said he definitely should not be driving, that we should take some positive steps immediately to stop his driving. Curtis explained that you, he and R.C. had talked to Daddy in an attempt to restrict his driving activities, without success, and asked if he had any recommendation. He recommended that someone talk to him again and tell him that the Dr. had said that he should quit driving and if that did no good, that we could contact the state drivers testing officials & get them to call him in for a drivers test. Curtis & I talked it over & we agreed to follow the Dr's recommendation. I was elected, so I talked to Daddy Sunday and told him that the Dr. sent word for him to quit driving because he was too nervous. I told him I thought it would be okay to drive to M'sap to get his mail but that he shouldn't get on the hiway to go anyplace. I also told him that if he wanted to go to town to get R.C. or June to take him and that I would be living next door in 51 days and that I'd take him anyplace he wanted to go. He said, "Well, I'll just take this dang pick-up back to them guys at W'ford & see how much they'll give me back on it. I'll walk to the post office to get my mail and just fiddle around here." He didn't get a bit mad or anything. I hope he does what he promised but he may forget about it before he gets around to it. In the meantime R.C. was going yesterday & talk to a friend of his on the State Hiway Patrol that he was sure would be sympathetic to our problems, and lay the groundwork for his drivers test in the event that he does ignore my advice. I hate like hell to do anything behind his back and I sure wouldn't have agreed to doing it in this manner, except that I thought it would hurt Daddy less, and be more effective, if someone in an official capacity did it than it would for one of us to do it – even if we could have.'

"When the Dr. told us that there was nothing that could be done to stop or help his condition Curtis told him that he didn't doubt his capabilities or those of any of his staff, but that his main interest was to help Daddy if at all possible, and asked him if he thought there was a possibility of error in their diagnosis and if he would recommend that we try another Dr. or clinic just to rule out the possibility of error. This didn't seem to upset the Dr. too much. He said there was always a possibility of error, and that he hoped that their diagnosis was wrong, but he didn't think so because, quote: "We're pretty good." He went on to say that he would cooperate with us in any way in the event we did decide to take him someplace else or if we wanted to bring him back there for further study. He recommended two Drs., one at the Timberlawn Clinic in Dallas and one in the Baylor School of Medicine in Houston. Curtis wanted to talk to all the rest of the family & get them to think it over & let him have their opinion on whether it would be wise. I told everybody but Bill about it yest (he wasn't there but is coming down here this week). I've been thinking about it all weekend but have not arrived at a definite conclusion as yet. I agree with Curtis on doing everything we can to help him; however there are many other considerations that must be carefully weighed as to whether such action would in fact help him or hinder him. I want go any further into the subject because I don't want to influence your thinking one way or another. As I said before, I haven't made up my mind yet.'

I think I have just about covered everything except the fact that yesterday was the most pleasant day I have spent at M'sap in quite a number of years. Daddy was calmer and wasn't "carrying on" with everybody as much as he did. Some, but not as much. Mama or anyone else didn't say one cross word to him all day. Mama & daddy was joking with each other and seemed closer than they have in years. One example that I sort of got a charge out of was: When we first arrived Daddy got up & said "I'd better get these whiskers whacked off before "Maw" gets me". Mama said "Oh Roy those whiskers don't bother me – you never kiss me anymore anyway." – and Daddy said "Ow! Maw!" and went into the bathroom laughing & shaved. There was still a bit of tension in the air when everybody was there, or maybe I was just holding my breath, but I feel things are definitely improved as far as our family relationship is concerned. I may find otherwise when I get up there and more closely observe, but I certainly hope not.'

"I really got a bang out of that Bryan every time we see him. Yesterday he "captured' him a horned frog and it discharged some blood on him & he really got excited because he didn't know they could do that. I don't know what happened to it but I think David must have shot it with a BB gun because he caught another one & came in the house to see if Mamaw had something he could put it in and while he was in there one of the other young'uns came in & told him that David wanted to see it & Bryan said "Yeah, he wants to shoot it," so I went outside to give David the word. Mamaw gave Bryan a fruit jar for the HF & he came on out and I asked him if he planned on keeping it. He said yes so I asked what he was going to feed it. He said he planned on letting it loose every day so it could find something to eat. I asked him if he thought it would come back after he let it loose and he replied "Oh, I'm going to walk with him while he's eating and when he gets full I'll put him back in the jar." He talks like a grown-up all the time.'

"I'll bet David & Chip ran 50 miles through the woods with the B.B. Guns. David was flaked-out when we started home. I'll be glad when we get up there where David can get out & do a few more things. There's not much he can do in a 50X100 ft. lot and his school activities have kept him tied up so late each day and all day on Sat. that I haven't been able to get him out for much outdoor life.'

"I guess I'd better stop before I have to send this by slow-boat.

"Take care & let us hear from you often."

Love from all,
(Loyd W)

Those of us at home tried to continue a normal life. I was in my second year as a school district superintendent and taking courses toward my doctorate. Shirlene, my wife, was a full-time mom and Matt was in first grade. Scott and Joe were at home, developing into robust kids. We were worried about Daddy and we watched the newscasts every evening, wondering how the North Vietnamese and Viet Cong could sustain an effective resistance after all the body counts were broadcast. Our superior forces should win this conflict soon and we could resume our lives without war worries. Life doesn't conveniently spread our problems over time, but at times offers them in bunches. Again Roy needed to write to Loyd. In part, this is what he wrote May 6, 1967. He had concerns about his safety, the welfare of Pat and the children, but additional worries about the deteriorating condition of Daddy.

"I had planned to do a lot of letter writing last Sun & Mon when I was on alert over at Nakhon Phanom but as it turned out I was in the airplane most of the time. We've been quite busy lately-mostly with our SAR mission. Seems that the NVN troops didn't get the word from McNamara that we're not suffering heavy losses. Also, they take exception to our 'undisputed air superiority.' In the skies the MIGS are getting more bold all the time and our people are avoiding contact a lot of the time. We sure as hell need something-wish I knew what.'

"As for our operation, we're still in pretty good shape-some of our rescue efforts are getting closer and closer to Hanoi but we've insisted on and in most cases received adequate MIG cap. We lost an A-1 the other day (May 19) on a royal screw up of an effort up close to Hanoi. [This date had to be an error, but the date would become very significant.] We were lucky that it wasn't more. Migs shot one down & swarmed all over another-he finally got away when some F105's showed up & divided their (the Migs) attention. Since then they've been paying a little more attention to where they commit us and what type of cover we have."

Roy continued to perform his duties in a very professional manner. We continued to hold the fort in Texas and wait until the day he could come home. He found himself in another rescue mission on May 14. His team was successful as reflected in the following quote from his citation for The Distinguished Flying Cross.

"Major Roy A. Knight Jr. distinguished himself by extra-ordinary achievement while participating in aerial flight as an A1-E Skyraider Pilot in Southeast Asia on May 14, 1967. On that date, while flying as a high element lead consisting of four A1Es and two HH-3C helicopters, Major Knight conducted search and rescue operations that resulted in the successful recovery of a downed pilot. Major Knight led the rescue force through rugged mountainous terrain, poor visibility and extremely intense and accurate antiaircraft fire to the area of the downed pilot. He successfully located the downed pilot and directed the helicopters in for the rescue under extremely adverse conditions. The professional competence, aerial skill and devotion to duty displayed by Major Knight reflect great credit upon himself and the United States Air Force."

We will never know precisely what happened in the cockpit of our brother five days later on a bombing mission to a North Vietnamese missile site in Laos. We know that we lost him that day and we know that he left this world as he had lived all of his life, courageous and devoted to duty. The citation for the Air Force Cross that was awarded for his actions that day, reads:

"Major Roy A. Knight, Jr. distinguished himself by extraordinary heroism in connection with military operations against hostile force as an A1-E; Skyraider pilot in Southeast Asia on 19 May 1967. On that date, Major Knight led his flight in a strike against one of the most important and heavily defended target complexes in Southeast Asia. Against overwhelming odds Major Knight pressed his attack. His aircraft was struck, resulting in loss of control. Major Knight, fully realizing that he could not regain control, jettisoned his ordnance on the target in a valiant attempt to destroy it and his aircraft subsequently impacted in the target area. Major Knight's unparalleled bravery and courage against virtually insurmountable obstacles were in the highest traditions of the military service. Through his extraordinary heroism, superb airmanship, and aggressiveness, Major Knight reflected the highest credit upon himself and the United States Air Force."

We are still waiting for a definitive answer to the question of his fate. Perhaps someday we will receive word that his remains have been found. Until then or forever we will wait like so many before and those who will come after us. But we know this for certain; Jack was right about Roy when he said he was tough. He will never be forgotten.

Roy

Junior was my brother, my friend.
He teased me and protected me,
His little naïve and gullible gnat.
He bought me a bike.
He encouraged me in my quest
To be what I could be and did
To show courage in my life's format

Roy loved his little girl Patricia,
Who became the mother of three
Precious and beloved children, so young
So innocent, so talent laden.
Too much so to be abandoned for dreadful duty
Yet too dedicated and courageous to flee
From the call of country and liberty.

Dad, husband, brother is still sought
By a government on a pace that would
Embarrass a sloth.
Seeking bodies of men who preserved our freedom
Has slipped from urgency
To a bewildering morass of tedium.
Why!
Will anyone help?

All that we know about the efforts of our government is contained in the following report disseminated at a family conference for MIA:

"In Sep. 82, a delegation from the National League of Families visited Vieng Sai, Laos. The delegation was accompanied by Col Khamla Keophythoune, MOI, who told the delegation that he had been a personal witness to the shoot-down of a radially piston-driven aircraft. He went to the crash site and saw a large body' the pilot was dead. He offered no further details on the disposition of the body. Col Khamla escorted the delegation to this crash site which was subsequently correlated to this case...(ANALYST NOTE: During a 18 December 1993 interview, Col Khamla denied ever seeing any dead American pilots, claiming he only saw a picture.)'

"6 Mar 85 (T85-038). A refugee stated that in 1980, while on a work detail near the Vieng Sai Airfield (VH 1958), he observed the wreckage of a U.S. "Skyraider." The site was located approximately 300 meters from the north end of Vieng Sai Airfield. Although partially scavenged, the frame, engine, and a four blade propeller were reportedly still there. Source further related that the aircraft was shot down by a PAVN AAA unit firing from Basn Nakay Neua (VH1957). Source heard conflicting stories regarding whether the pilot was captured or killed. ANALYST COMMENT: A 1994 excavation of a crash site exclusively correlated to this case confirmed that Maj Knight was in the aircraft at impact.'

"2 Feb 87 (T86-580). A refugee reported seeing partial aircraft wreckage in the area of Vieng Sai Airfield during his 1978-1985 seminar confinement. This report is very similar to T85-038. Source did not know when the aircraft crashed, nor did he know if anyone had died in the crash. He said that the crash site area had been cultivated and extensively built up since the war.'

"22 Dec 87 (T87-496). A refugee stated that while confined to a Bieng Sai seminar camp from 1975-1987, he had heard from an unnamed former pathet Lao officer that the pilot of a crashed "Skyraider" aircraft had been buried about 200 meters east of the north end of the Vieng Sai Airfield runway. The grave was said to be near the bottom of a small rise about ten meters high. Source also recalled with great detail the 1982 U.S. delegation (National League of Families) visit.'

Gayann attended this conference and was told that the reason this information was not further investigated was that only ten cases per year were being worked. When she asked for a reason for this, no answer was forthcoming. Why?

As I was writing this chapter, I speculated as I talked with Gayann about what might have happened in his mind and the cockpit at the time he was hit. With very little understanding of the conditions and the training of pilots, but knowing Roy, I voiced the possibility that he might have intentionally flown his plane into the enemy position, if he was unable to eject or right his plane. She passed this theory along to Bryan. He immediately wrote me with his more informed opinion in a letter dated May 9, 2002,

... "On my way to the airport yesterday I got to talk to Gayann a little. She mentioned you've been working on your book. That is so great. One thing I wish I had some talent for is writing.'

"She said that Curtis has, lately opened up some and told more about Burma. I can only imagine the images that are in his head. What it must have been like for him. What it must have been like for all of them. It just reinforces my belief that, although I was a "professional soldier" for 14 years and dedicated my life to preparing for it, war is absurd.'

"Gayann mentioned that you had considered the possibility that in Dad's crash, he might have tried to crash his aircraft into the gun emplacement. From what I know of my father and of his brothers I have to say he was a hero. But I will say that having had almost the exact same training he had, he was more than anything a very highly trained combat pilot. That means that his first job was survival, his second was the mission. I can say with some degree of confidence that if he had any control of the aircraft at all he would have tried to recover it and return. If he was alive and capable, the next thing he would have done is eject from the airplane.'

"From reading his shootdown description and having had a little experience in an out of control situation, here's what I believe happened.'

"When Dad was first hit, he knew he was hit and he did exactly what he was trained to do. He jettisoned his ordnance. This makes the aircraft much lighter and more stable and would help in recovery. At this point his wingman said his plane tumbled. One of two things happened I believe. First, he could have been injured by additional AAA, such that he was unable to eject. When they fire AAA, they fire in volleys and continue firing. Second, the tumbling aircraft could have inflicted acceleration (g's we call it) such that he was unable to eject or was rendered unconscious.'

"It's not that I don't think Dad would do something like that. I just know that he was highly trained and dedicated and his shootdown was consistent with what would have happened. It is amazing though, having seen examples myself, the courage people will display at times, often the people you don't expect.'

"I just thought I would pass on what I thought happened since I have a unique perspective considering my own experiences and training. Of course no one really knows for sure."

I asked Gayann to send me some information about life with her Dad and family. She did much more. The following passage is a poignant memory search of a little girl about her beloved Dad exactly as she wrote it.

"Dear Bill,

"Well, I haven't had any good quality time to sit down & write what I remember about Dad. I am sitting outside a Travis County courtroom waiting to be empanelled on a jury. I'm going to try to get out of it, even though I would be a great juror. I watch a lot of "Law and Order."

"I have been thinking about all the things I remember about Dad. The very first memory was a scary one. It was scary because it was very noisy. I must have been around 2-21/2 & we were at an air show – go figure. I remember the loud jet engines & I was scared to death. Dad picked me up & held me. I remember him laughing & holding me tight. I was still scared, but felt a little better because I knew he would keep anything from hurting me.'

"I remember Europe – kid memories, simple memories. I remember France because of the French bread – that's me – always the chow hound. I remember the cobblestone streets & all the French people. One day, Dad, Chip & Bryan & I went on a walk in the woods nearby. The trees were real big & thick. It was like a dense forest as opposed to Texas scrub land w/ a few tree clumps. This was your basic deep forest ecosystem. We were looking for snails. We found 'em. They were big giant bright orange slugs. They were big as half your hand. I really remember how he gathered us all around & showed us these nasty looking snails. I also remember Dad instructing Chip & Bryan on how to properly pee on a tree – they had to go.'

"Most of the time in Europe seemed to be spent with Mom a lot. Dad always seemed to be gone flying – TDYs.'

"I remember getting burned somehow – I think in France. I don't know how, but I remember having to have bandages soaked & pulled off. I can still see Dad leaning over the bath tub w/a cigarette hanging out of the corner of his mouth, eyes squinting, gently pulling stuck bandages off me. He was always take charge and in control.'

"The big boat ride back to the States was fun. It seemed to take forever & we kept puking & wetting the bed. I remember running through the ship with Chip. Bryan was probably less than 2 & close to Mom all the time. I remember Chip & I chasing down a rumor that they were showing a Roy Rogers movie somewhere on the ship. I can remember going down long hallways on the ship trying to find that Roy Rogers movie – never did.'

"Del Rio was the beginning of the very vivid memories for me. I remember living in this cheap little carport apartment when we first got there. I think I was 4. Bryan was too little to really hang with – other than from the "baby to play with" perspective. I can remember getting stung by a bee while I played in the back yard. I seemed to always be waiting for Daddy to come home. We finally moved into a house in Del Rio & our Air Force roots were finally set down – just for a little while.'

"This is when I really seared that vision of Dad in his flight suit. He came home everyday in that flight suit. He would go to the freezer and get the polar bear whiskey out of the freezer & pour a drink in those fancy glasses. The best part was when he put the olives or cocktail onions in. We loved those and got in trouble regularly for eating them out of the jar in fridge door. The "polar bear whiskey" was vodka I believe. Each bottle came with a little plastic polar bear on a chain – like a key chain. We 3 kids would always fight over it. Believe me. There were plenty of little polar bears to go around. Dad would make runs to the border to pick up alcohol real cheap. His big drink was a martini – hence the previous description. I can see him sitting on the couch, with an ice cold martini and a cigarette – unwinding from the day.'

"One thing about Dad – he was never lazy. He was constantly doing something. He was exceptional with his woodworking & car-working skills. We always had scraps of wood, lots of tools and at least one car he was working on.

I can remember that yellow & white – I think, '57 Chevy? Then…He decided he wanted to refurbish a Ford Model A. I can remember him finding an old shell of a Model A –in Mexico – I think. I remember when it was towed up in front of the house. It was a junk heap. Rusty, dirty, long ago abandoned. I remember thinking that he didn't get a very good deal to get such an ugly car. He spent the next at least 3 years refurbishing the car. It was amazing how hard he worked on that thing and how much he loved doing it. He worked on it all the time.' [Bryan, his son, carries on the family tradition for 'tinkering'. He is currently assembling a jet aircraft in his garage. I suppose this is a natural transition since he has an engineering degree and twenty years experience flying jet fighters and commercial aircraft.]

"My Dad was the greatest Dad. He played with us, loved us, talked to us, took us fishing & to family reunions, helped us discover things, taught us honesty and honor and to always do your best. He was handsome, intelligent, extremely funny, very resourceful, a good provider and a mountain of strength. That was my reality until I turned nine. My life was good, secure, balanced and full of love. That changed!'

"I can remember the exact moment that life began to change & die – never to be experienced again. I was playing outside with Chip, Bryan & some neighborhood kids. We were having a great time. We must have been playing chase or something that required running fast because I was winning & having a great time. It was a glorious, beautiful day. We were interrupted in our childhood frolicking by Mom who called us three in the house for a few minutes. We all ran in, sweaty & breathing heavy. I can remember being a bit annoyed that our play was being interrupted. I remember sitting in the living room side by side on the couch. Mom & Dad proceeded to tell us that Dad was going to Vietnam. Vaguely I even knew what that was – something we heard mentioned on the news and unless, it was animated, and had Bugs or Johnny Quest in it – we didn't pay much attention. I remember thinking what's the big deal. Is this like a TDY? Can we go back outside and play? I had no concept whatsoever about what it meant. I remember kind of blowing it off – actual words, but I noticed the looks on Mom & Dad's faces. They looked sad & concerned & serious. The word Vietnam was not synonymous with death for me just yet. I remember being real happy when they let us go back out & play, but I kept wondering what the big deal was about this "Vietnam" thing.'

"The next few months were wild & crazy. We assumed the position for another move. We were getting good at it. The Air force movers would come in, wrap everything we owned from bobby pins to shoes in beige paper. They would stick ID tags on every piece of furniture we had. They were efficient & quick and in no time we were once again living a day or two in an empty house until BO Qs could be arranged. I can remember our last day at Laughlin AFB. Chip, Bryan and I were running around the completely empty house. It was early in the morning. Mom was there, but Dad was gone. We were running around being loud & obnoxious when in walks Dad with a bunch of doughnuts, milk & juice. It was similar to a momma cat bringing home a dead squirrel for the kittens to eat. We devoured those doughnuts & then we started on our next journey.'

"I remember, we lost our cat Tarbaby when we moved to the BOQs. She disappeared for a couple of days & we thought we lost her for good. Dad went back to the vacant house one more time before we left for Florida & guess what? Tarbaby had gone back to the house. Plucky cat! Dad brought her back & I was once again a happy little girl.'

"The next week was spent traveling to Florida. We were towing the Model A. Dad wanted to finish it before he left for Vietnam. I can remember driving & driving & staying in hotels & eating a lot of shrimp along the way. It was like my main meal. I guess Dad could afford it. He never told me not to order shrimp.'

"When we arrived in Florida, we were set up in a fairly large apartment complex. The Air Force had pretty much dominated that complex with families like ours. So, we had a lot of built in Air Force families. The complex was designed so that there was a big almost T-shaped courtyard in the center. The parents/pilots would sit out in the courtyard and drink & talk. The kids would play all kinds of games. One of the favorites we played was swinging statues. This game involves one kid swinging another by the arm around & around. Then the kid is let go & falls. The kid freezes in the same position. Then it is up to the swinger to guess what the kid is a statue of. I can remember playing this game in soft, fresh, green grass while our parents talked, laughed & drank. Most of the Dads wore the same green flight suits.'

"There was a difference with this location that was a bit different than the others. There seemed to be a tension there. Mom was real stressed & bitching a lot. She seemed worried & not real happy. Some of the other parents seemed tense too. Now I understand, I was in 3rd grade and was not very experienced with life or people, but I could feel an uneasiness with all of the adults. It was like they were pre-occupied. It was like they couldn't have fun or really relax or be happy. It was like they were waiting for a bad tornado to hit.'

"Those days in Florida were filled with school, sand, a three-legged dog & a fort made out of Yucca leaves. Dad went to work everyday in his flight suit. I don't remember any TDYs in Fl. I think Fl was the TDY. The climate, trees, bugs & people were much different than Texas. The humidity was stifling, but the Spanish moss that hung off the Cypress trees was big & beautiful. It always felt like the trees were big umbrellas with big play areas underneath.'

"Batman was real big at that time with the TV show. I will never forget 5 year old BK [Bryan] laying on his stomach on a swing with a towel safety pinned around his neck singing...na na na na na na na na na na na na na na na na BAT MAN! The whole time he is swinging back & forth on his stomach like he is flying. So cute.'

"I remember following my Dad up the apartment stairs one day when he got home from work. He had white salt stains from sweating profusely on his flight suit. I told him that his flight suit didn't get cleaned real good last time cause he had white stains all over it. What were those white stains. He laughed and said it was sweat. OOOKY!! Lots of sweat. Then I started thinking about how hot he must have gotten to sweat so much. I felt sad & worried that he got too hot at work. That's when I started thinking his job must be real hard to get so hot & sweat so much. I had a new appreciation.'

"One of the best memories Chip, Bryan and I all share is the building of the fort in Florida. There was an undeveloped greenbelt across the street from the apartments where we lived. Dad took us all out there one weekend and saw the potential of all the yucca type palm leaves lying around. I'm sure the original mission was to go exploring in the woods. Soon, we were gathering up all these leaves and Dad started building this fort. It was the coolest fort. It was short, so Dad would have to scrunch down to get in it. But, it was the right size for all of us kids. We trekked out there a couple of more times to finish our fort. It got to a real good shape before it was finally abandoned. It was so much fun. He was such a woodsman. He always loved fishing and going out in the woods to explore.'

"Of course one of the most vivid memories I still have is the time I got the fish fin stuck in my foot out at Mamaw's and Papaw's. You probably remember it. I think ya'll were there. It seemed like everyone was there. It was an Easter weekend, I think. All the Knight Brothers were fishing at the tank. All their kids (us) were running around getting into stuff. The situation was that Dad and the others were catching mud cats. When they caught them, they threw them back over the little hill. Dad, being the astute parent he was, called all us little kids over and showed us these mud cats. He held it up to us and said, "Ya'll stay away from these fish. They are bad fish." Well, being the oldest of my pack – Melody, Bubber, Bryan and me, I headed up the group to go over and kill those "bad fish". So, our fearsome foursome went over to the fish and started stomping them as hard as we could. Of course, within about 20 seconds, I stomped a mud cat fish thorn straight through the center of my foot – all the way through. Well, the second stomp after that fish fin was set – really hurt! I knew something was wrong, but I had no idea what. My instinct was to get back to the house where Mommy was. The trouble was, I couldn't walk very well. I hollered at Melody to come over and help me. I put one arm over her shoulder and tried to walk back to the house. I needed another person to lean on – Bubber was closest, so I grabbed him. Trouble was, he was too short. Then I got Bryan. I hobbled to the house and started crying right as I got to the back gate. Of course as soon as the women folk saw what was gong on, all hell broke loose. Someone picked me up and carried me into Mamaw's kitchen and set me in a chair. Someone told Melody, Bubber and Bryan to run to the tank to get Dad. While we were waiting for Dad, they tried to take my tennis shoe off. That didn't work. The fin was through my shoe and through my foot like a nail holding up a cabinet. Finally, my Dad showed up and took charge. He pushed the women out of the way, summoned a razor blade and went to work. He cut my tennis shoes straight down the middle front so he could peel it off my foot. It worked like a charm. Once he got the shoe off, he picked me up and took me to Camp Wolters to get the fin removed. It was just him and me. Or at least, I just remember him with me We got to the Medical Center and he picked me up to carry me in, while he was doing this, he accidentally burned me with his cigarette. I didn't mind. Before too long, a doctor was pulling the fin out of my foot with hemostats. Waaalaaa! Fish fin is out, tetanus shot is in. They wrapped up my foot and were on our way back to Millsap. My Daddy came to my rescue and took care of me. When we got back to Millsap, I was banned from the tank. Typical women! O misery of all misery. I had to limp around for another day and a half in the house while everyone else was down at the tank fishing and having a blast. I was not very happy. I carried that fish fin around for a long time until I finally lost it. It was about an inch and a half and I took all of it in the foot.'

"One of our memories as kids was when we were at some place in Southwest Texas. It was when were living in Del Rio. For some reason I want to say it was Eagle Lake or Eagle Pass. There was also a river. It was a red river with moss over everything. It was extremely slippery and deep in some spots. We were all splashing around, but because it was a river, there was a current. For some reason, Dad had to get us 3 rug rats across the river or down to another spot. So, he comes over to us and tells us all to hang onto his arms. He took both of his arms and bent them at a 90 degree bend. We all grabbed onto his forearms and hung there like big sacks of potatoes. The water was a little higher than waist high on him and we were all dangling in the water hanging onto his forearms. We all started talking in childlike amazement that he could carry all three of us at the same time and carry us downstream. We were all so amazed at how strong our Daddy was. He was the biggest, tallest, strongest man on Earth and he was our Dad. We're so lucky. We talked about that for years afterward. We would tell our friends about how strong our Dad was and then describe how he carried us in the river. Everyone was always amazed at that story. Of course, none of us were figuring in the buoyancy of our bodies and how he wasn't carrying that much weight. He was just the strongest Daddy in the whole world.'

"We spent our last Christmas as a family in Florida in 1966. I was 8 years old. We had a little 2 bedroom apartment. The boys and I stayed in the same room. We had a tiny living room where we put a smaller than usual tree. It was still magical and there were plenty of presents to go around. We had been told that Santa knew we were moving to Millsap in January and he was delivering our presents there. Santa was so smart.'

"I remember what I got my Dad for Christmas that year. I spent years dwelling on what a worthless, meaningless gift it was when we were all facing such an uncertain future. I just didn't know there was a possibility that he might not come home. This was just another TDY. I remember not being able to find anything I liked at this particular store where my Mom and I were shopping. She was getting impatient and so I had to choose quickly. I picked a compact, travel size shoeshine kit for my Dad's last Christmas. What a goof. I just had no idea he wouldn't come back.'

"We have that last Christmas on video. It is real cute and so sad – now.

"I believe we kicked ya'll out of the house in Millsap when we got back, right? [No. We had moved in July 1965 to Saltillo for my first superintendent's job.] Well, we set up house real quick. I don't know how long Dad was there before he left – maybe a week or two. It seemed like we did a lot in that time.'

"It finally came that Dad had to leave. Mom and Dad had decided they would keep everything as 'Normal' as possible which meant Chip and I went to school. I really wish we had been able to go to the airport. The night before he left, he went around to each one of us privately and said goodbye. That was the first inkling that this was serious. He never did this at other TDYs. I only saw him saying bye to Chip. It was a small house and I was sleeping in the living room on the couch. My Dad was hugging Chip in the hallway where those two rooms were separated by the bathroom. Light was illuminating from somewhere – maybe the bathroom, but they were in the dark. I really saw silhouettes of Dad and Chip hugging and crying. Chip, by this time, knew what was going on. He knew Dad might not come back. That's when I started thinking it was serious. Did everyone know something I didn't?'

"After Dad said goodbye to Chip, he came into the living room where I was in the dark. I was in my jimmies and I had a blanket and pillow on the couch. He came over and sat down beside me and basically told me to be a good girl and do what my Momma told me to do. He told me he loved me, hugged me, tucked me in and said, "Goodnight, Honey. I didn't know enough to cry. I was just very aware and puzzled by the intensity and seriousness of everything.'

"I didn't see my Dad the next morning. He must have stayed up late – he was still sleeping. Chip and I got ready for school, caught the bus and started our life without our Dad.'

"My Mom has the airport story. I just have a picture of them going to the airport (Love Field) – very somber. Bryan was with them. I know she cried a lot. I don't think Bryan has a memory of it. He was only 5 at the time. His memory is not as good as mine.'

"That was it. He was gone.'

"This was a very terrible memory for me. It was a Friday night. I knew this because Chip had gone with Soup and maybe you (and a bunch of tag alongs) on a fishing camping weekend. It was just me, Mom and Bryan. I remember I had gone to bed. Sometime in the night I woke up because a light was shining in my face. It was the kitchen light. I heard movement. I could also see into Chip's room and he was there. I decided I needed to investigate. I walked into Chip's room to make sure I was seeing what I was seeing. Sure enough, there was Chip, fully clothed, shoes on, asleep like someone had just laid him there.'

"I then went into the kitchen. There was my mother sweeping the floor. The kitchen had been a terrible mess the night before. Mom never was a good housekeeper. Maybe that's where I get it. Anyway, I went up to her and asked why Chip was home. She stopped sweeping and said, "Your Dad has been shot down.' I said,"Oh." I turned and went back and got into bed and laid there wondering what that meant.'

"The next morning, people started coming. They brought food, cakes, and pies. People I didn't even know. Everyone was crammed into that tiny house. My Mom was always sitting at that tiny table in the kitchen. Bryan seemed to be on her lap a lot. I think the story goes that Mom was crying a lot and Bryan was following her around to make sure she got enough Kleenex. I remember watching Bugs Bunny/Road Runner cartoons with Vicky and Ronnie. Chip and I were in the room. Evidently, Lorene had given them instructions to stay close to us and keep our minds busy. Ronnie with Chip, Vicky with me.'

"Side note, Ronnie really loved my Dad. I think he almost idolized him. I can remember him telling stories about my Dad and his fishing skills and calling him N..... and Junior. Dad gave his very best fishing worm to Ronnie right before he left. I can remember him trying to decide who to give it to. I think because Dad and Ronnie both were such outdoorsmen, they had a kindred spirit. Anyway, I'll never forget when Dad gave it to him. I was really a little fisher girl myself. Whenever Dad was fishing, if it was remotely possible, I was there – usually just messing around, till I got bigger and could fish. I remember Dad looking in his tackle and digging around for that tattered, smallish red, rubber worm. He pulled it out and said that he thought Ronnie ought to borrow this worm while he was gone so it wouldn't be idle. You would have thought you had just given Ronnie a spiffed up motorcycle. He cut a great big smile and said, "Thanks, Junior. I'll be sure and catch a bunch of fish with this. Don't worry about me taking good care of it"...and on and on. Evidently, Ronnie knew about this worm and was just flabbergasted by how many fish it had caught.'

"Everything was going fine. We were waiting to hear if there had been a parachute. Finally in the late morning, some Priest or Chaplain came to the house and gave us the bad news. No parachute. As soon as that was announced, Lee started wailing and had to be carried away – I think. I remember I was mad at her for a long time for scaring me so bad and bringing punctuation to such horrible news. As soon as the wailing started, I got up from the cartoons and ran into my Mom and my room. I slammed the door, threw myself on the bed and started crying real hard into the pillow. A few minutes later, Lorene came in and didn't say a word. She just sat with me and rubbed my back while I cried myself to sleep. I always loved her for that act alone. I knew someone was there who cared for me – relative. I knew she was crying too. But she was tough and strong and no one else came. [Lorene had a brother shot down during World War II and was never found.]

"Within a day or two, people quit coming and the food stopped. I remember my Uncle Gary came 2-3 days later. He was in the army. I can't remember if he was stationed at Fort Sill or Fort Hood. I want to say Fort Hood. Anyway he was goofy with us kids, kept us busy and thinking and laughing. Then he left. And you know the rest.'

"I have a bunch more little memories of my Dad at different times. But, I'm kind of worn out from this project. I always keep my Dad's memory deep inside me. I let the memories out at different times, but in small quantities. Otherwise, I get all depressed and cry a lot."

Gay, thank you for having the courage to share your memories.

The absurd and ambiguous war lasted another eight years, giving Herchel's son time to graduate high school and the University of Texas at Austin. After he earned his bachelor's degree he entered the U.S. Air Force in 1973, becoming an officer. Ironically, he served in Southeast Asia from March to December 1975, flying out of Udorn, Thailand. His only action in the Vietnam War was on the day of the evacuation of the American Embassy. He was Weapons Systems Officer on an F-4. He and his pilot were flying with a group of four F-4s, providing cover for the helicopters that were transporting the evacuees. He told me that he was in the fourth plane and spent most of the time looking behind him for MIGS. My criticism of the Viet Nam War has nothing to do with the men who fought it, but with those in high places who ran it. My admiration for the warriors is deep and everlasting.

Roy's military legacy continued through his youngest son, Bryan, who was six years old at the time Roy was shot down. He received a degree in engineering. Through the Air Force ROTC he was commissioned a second lieutenant upon graduation. Bryan went through flight school and flew various aircraft and while serving as a pilot instructor at Sheppard AFB in Wichita Falls, Texas he missed the Gulf War. He would later have the opportunity to demonstrate his courage and his skills as a pilot.

The warplane, approaching the reviewing stand from the left, was dark and menacing, a bundle of the most advanced technology in the United States arsenal. Much like a primeval avian predator, it swept by the stands and arrogantly sliced into an upward arc away from this Sunday afternoon air show crowd. Then the unthinkable happened. A piece of the left wing from the $43,000,000 top-secret aircraft broke away and the stealth fighter pitched to the left and assumed the nature of an autumn leaf falling to the ground. The pilot was pinned to the instrument panel by the G force, unable to move for the controls. He tried vainly to use his finely honed skills to gain control. A sudden shift in the plane's momentum freed him.

Major Bryan Knight tried, but couldn't control his aircraft. He was trying to guide it away from the thickly populated neighborhood below to a nearby lake. The F117 went nose down. His few seconds for action had passed. He ejected.

Miraculously he was shot free of the helpless mass. His parachute appeared to the applause of the crowd and he floated to the ground while observers caught the action on video. Bryan's parachute carried him through the smoke from the burning wreckage to a landing in this quiet Maryland neighborhood among the trees in a back yard only a few yards from the fireball created by the crashed plane and the house it hit. All of this action came within a few seconds. No one was injured, not even Bryan. This happened September 14, 1997.

The neighbors ran to the site to help Bryan. He and the folks in the community of Bowley's Quarters gained a few days of notoriety as the accident was carried on national television. The Secretary of Defense, William Cohen, spoke of his courage, skill and dedication. Asked if he considered himself a hero, he said that he didn't. Bryan thought of himself as a professional warrior doing a job that he had been trained to do. His actions were courageous, but were born of self-preservation, made necessary from careless construction and maintenance. A few bolts, worth perhaps five dollars, had been left out of a panel on the left wing.

The year before, on June 25, 1996 he reacted in a way that brought him recognition as a hero. He chose to risk his life for his fellow airmen at the terrorist attack on the Kobar Towers in Saudi Arabia. At the presentation of the Air Force Commendation Medal with valor device to Bryan, his citation read in part,

"Immediately after the terrorist bombing of Kobar Towers, Captain Knight went to Building 131, discovered a damaged medical vehicle near the crater and removed all medical supplies. With these resources he began to treat the severe wounds of several personnel. He directed ambulances as they arrived on the scene to insure that injured members received treatment. He then began a search in the rubble of the dormitory where he found three airmen buried underneath the debris. One of the individuals was still breathing and Captain Knight gave immediate first aid to stop the bleeding and carried the individual to a waiting ambulance. Another member had lost large amounts of blood and although the airman was not responding, Captain Knight continued to use his medical expertise to reduce the bleeding until the arrival of medical personnel. Captain Knight clearly saved the lives of several Air Force members. By his prompt and humanitarian regard for his fellow man, Captain Knight has reflected great credit upon himself and the United States Air Force."

Bryan was thirty-six years old, the same as his father during his brief tour of duty in Southeast Asia in 1967. Was he emulating the actions of his father? Were his father's heroic deeds as a pilot fighting the North Vietnamese affected by the actions of his older brothers in Burma twenty-two years earlier? How much was all of this the random results of a marriage at a time and place unique in history and of those inherent traits being forged by circumstances into a passion for dedicated duty?

As we look at the great generation that brought us through the Great Depression and World War II, it is easy to see how life's struggles in one prepared us for the challenges in the other. Could our victory in World War II have been as expedient without the toughening we acquired in the economic disaster of the nineteen thirties? Not likely.

We were later softened by improved economic conditions and the malaise of the post-reinforcement pause after victory over Germany and Japan. The Korean War was fought virtually by the same generation involved in the war just concluded. Men like Roy, a product of the Great Depression and World War II carried the roots of character and honor with them to the Viet Nam War and excelled as warriors, notwithstanding the absurdity of the commitment and strategy employed. He and hundreds of thousands fought on in spite of this betrayal.

Bryan had no personal experience in economic hardship. He never had to bend and sweat to help earn a place at the table. He never had to be embarrassed by his clothes or the family car. Yet, he delivered the same behavior seen in his father and uncles. His hardship was his legacy of living his childhood and adolescence, those formative years, without the strong leadership and love of his father. His salvation was the example of history written by his honorable father and to a lesser extent, by his uncles (including his mother's brother who fought in Vietnam). Even more significant was his genetic disposition through a long lineage of tough, courageous and honorable men and women. His behavior and character exhibited in our modern, technological world reflects the very nature of this tale as it began in 1915 with the start of Roy and Martha Knight's life together as it grew through the convulsive years of the twentieth century.

CHAPTER 6

WARRIORS GETTING READY

GOING BACK TO THE F Troopers, Jack's letter of August 5, 1941 discussed briefly his time at home for the annual homecoming at Holder's Chapel Methodist Church and his trip back to Fort Bliss. He referred to the homecoming as 'First Sunday in August', as we all did, because that was the established date for the reunion.

Holder's Chapel Methodist Church was located down the hill from Red Mud School. Mamma attended school at Red Mud through the seventh grade. At the turn of the century, small schools dotted the countryside. There was a school within reasonable walking distance of the farmhouses. The farms averaged eighty to a hundred acres in this farming region of Texas, leading to a well-populated farm community. She and her siblings walked the three miles to school when weather permitted and work in the fields didn't intervene.

The church building was located near the road to the school. The cemetery was located nearby and west of the church. The building was architecturally nondescript; a rectangular, utilitarian structure that served as little more than protection from the weather. It had taken the name of my mother's grandfather, Isaac Holder, who was instrumental in its establishment. A permanent arbor was built on the grounds as a solution to problems with the primitive brush arbor that needed constant upgrading. This structure was used in the summer for revivals and the annual homecoming. Anticipation for First Sunday in August was second only to Christmas as a holiday.

We had 'dinner-on-the-ground' which wasn't on the ground at all, but a pot-luck feast laid out on home-made, rough hewn pews pushed face to face. In the morning there was a modified church service with a mini-sermon, short testimonials and reminiscences and many of the favorite hymns sung. Daddy was one of the song leaders.

We called the noon meal dinner and the evening meal supper. So dinner was an elaborate layout of all the mammas' favorite recipes. I especially remember the fried chicken, potato salad, black-eyed peas and store-bought light bread. I can't forget the ever-present delicacy of deviled eggs. There was always a wonderful assortment of desserts, my favorite being banana puddin'. The birds used for the fried chicken could range from the more succulent young fryers to the tough old roosters that challenged the jaw strength of the younger kids. You soon learned to pick the smaller pieces. Hens were usually used for the awesome southern dishes of chicken 'n dumplings or chicken 'n dressing.

Barrels of water with huge chunks of ice floating in them were placed under the grove of post oaks next to the arbor. Tin cups on wire hooks were hung off the rim of the barrels. It was here that most of the school-age kids congregated to either play in the dirt, listen to the older kids tell their tall tales of sexual conquests or the more truthful tell their stories of frustration, or to watch the fights. This setting produced one of my favorite stories from the family folklore.

Since we were a large family with both parents from large families, we had a very abundant supply of cousins. Some of my Holder uncles and aunts had moved their families to West Texas to work in the oil fields or in search of larger or more fertile farms. Billy Holder was Uncle Bill's son. He had entered into a dispute with a boy from another family. Kinfolks didn't fight each other in public, especially at church gatherings. The contest was heated and provoked heightened passion from Billy's younger brother, 'Cush'. As coincidence would have it on this day, the best of the battle coincided with the close of the morning church service. At the exact moment that the adults under the arbor became silent to hear the benediction and food blessing, Cush was moved to inspire Billy with his outcry, "Whup his ass, Billy!"

Not only did his cheer reverberate through the arbor, it echoed throughout the valley. Uncle Bill quietly left the arbor, gathered the two culprits and headed for the thicker woods beyond the cemetery. I don't know and I doubt that anyone besides the three knows what happened in those woods. If Uncle Bill took his belt or tree limb to them, auditory observers wouldn't know, because the code didn't allow audible crying and their innate toughness wouldn't allow tears.

The Boys were waiting. After nine months of their 'year' of expected service, the prevailing attitude was to make it through the easiest way possible. Jack's letter of August 10, 1941 that was begun at Ft. Bliss and mailed in Louisiana alluded to this mind-set.

"Curtis is leaving with the 112th tonight and is going to San Antonio with them as they go to La. He got to go to the Cook and Baker School. He is the luckiest human I ever saw. He'll miss all the maneuver…Looks like we are going to be stuck for 18 more mo. doesn't it?"

Curtis wrote from his new location, Randolph Field with this description.

"Got here yesterday. The place is about 4 or 5 miles square. There are swimming pools and baseball fields all over the place. I am going to stay here 60 days and then go back to the troop at Ringgold. Lt. Hornsby said that I'd get $51.00 or $56.00 a month after I cook for the troop awhile. I can take off and go to town anytime I want to here. Don't have to get a pass. Don't work on Sunday. I cook for about 1,500 men. We have 50 cooks. I work one day and am off a day. I am really going to like this place. The kitchen is about 75 yards sq. Sure is cool. I never saw so many airplanes in my life. It's 18 miles from San Antonio, but there's a little town of our own out here at Randolph field. I haven't seen an officer since I've been here. Nothing but Sgts. and Pvts."

The Boys were not that far removed from the farm and could still be awed by their surroundings. I know that he had used swimming pools, but it was still considered a luxury. My only exposure to this privileged sort of life was at the Baker Hotel in Mineral Wells where we would climb the rock wall by its pool and gawk at the resorters while Daddy was peddling his produce inside.

They knew that they weren't getting out after a year of service. They knew they were going to get involved in the war and that it would last a long time, but they continued to speculate hopefully about their separation from service.

The war games in Louisiana in August and September of 1941 were designed to test the war readiness of U.S. troops and refine tactics and strategy. Even though the army had surely reached the decision that horse mounted cavalry was obsolete due to the decimation of Poland's crack cavalry regiments by Hitler's Panzer divisions the year before, they were included in these maneuvers. This was likely the result of lobbying from the proponents of horse cavalry, the last gasp to retain an age-old component of military history and the nostalgia and romance that it engendered.

In early August of 1941 Winston Churchill and President Roosevelt met on board the HMS Prince of Wales off the shore of Newfoundland. The conference led to an agreement that moved the two countries into an inextricable alignment. The signing of the Atlantic Charter committed the United States to the assistance of Great Britain and an emphasis on the war in Europe. The United States agreed to provide supplies and to focus on the war against the Axis powers. At the same time, Roosevelt's representative to The Soviet Union offered Stalin help to strengthen his stand against Germany. The nation's course was set. The Japanese continued to plot their invasion of Southeast Asia and islands in the Southwest Pacific. (WWII, American Heritage)

Jack, Curtis and Loyd were caught in this tide of world politics and against their plans would be swept along in its swell for years to come. How would they do in this effort? We can find some clues from their early years. My early memories of the triumvirate, The Boys, have surely been influenced by the stories passed to the younger children after they left home for college. Those stories were told at a time and in such a way that I internalized them as a message to us that we should do better. The images of the three that grew in my mind were different, but invariably positive.

Jack, the first, became the leader, the one that all of us deferred to, not without some challenges from Curtis. He began life as the prize of two doting young parents. Daddy told me of the time of his first paddling. He was four and had grown to expect to go with Daddy any time he left the house in his wagon. On a day that he was to be gone too long for Jack to accompany him,

Daddy told him that he couldn't go. Jack threw a typical childish fit for being denied. While Mamma restrained Jack, Daddy left. When he was down the lane about a hundred yards, he heard a noise and looked back. Jack was running toward the wagon, yelling and carrying his shoes and socks. He had broken away from Mamma or had been released. Over Mamma's protests, Daddy broke off a small switch from a roadside bush and gave him a few whacks.

Jack grew to respect firm discipline and helped in the control of the growing herd of boys. Mamma was very possessive and was reluctant to turn over control of her first born to other authority figures. Jack didn't start school until he was seven and was nineteen soon after graduating high school.

Jack served as straw boss for Daddy, especially for the gooseneck squad that had to do the grunt work of hoeing the grass and weeds from the crops. He not only had been delegated the authority, he also had the presence and strength to maintain it. With some college under his belt, his sagacity waxed and his sphere of influence expanded. Mamma and Daddy would come to value his advice and, in fact, seek his counsel.

While interviewing his grade school teacher, 'Miss Lou' Alice Kimbrough, 93, in June, 2001 she told me Jack was a serious student and the song he always wanted to sing at assembly was 'When the Roll is Called Up Yonder.' She also told me in her own irascible style that no matter what I accomplished in life that I would never reach Jack's character or accomplishments. I agreed on both counts without equivocation. On the same day I interviewed 'Miss Cloe' Fine Smith, 91, also his grade school teacher. She described Jack as all around great guy, mature, peaceful, not overbearing. She told me about Daddy complaining in a teasing way that June's birth had ruined his effort to build a baseball team. She said there was no one more beloved than Jack.

Curtis, being the second, thrived on the divided attention of his parents. A well-behaved, conscientious, hard-working boy, he showed an indomitable spirit. As a young boy, I was compared to Curtis and he became a model for me. While cultivating this notion, Mamma could get me to work extra hard in the garden, while sweeping the floor, hoeing in the field, or churning the cream for butter by bragging about how well Curtis did these things at my age. He was probably motivated by the desire to finish an unpleasant task or to grow strong enough to whip Jack. I never felt like I completely met the challenge, but I tried. I was never motivated to work that hard and never had any illusions about whipping Junior the Terrible, my brother just older.

All of The Boys had a well-developed sense of humor, probably acquired from association with Daddy who had a propensity for good-natured tomfoolery and story telling. Miss Lou described Curtis as feisty. Miss Cloe called him high tempered. When disciplined by having him stick his nose in a ring on the black board and the class laughed, he turned around and shouted, "There's an awful lot of cackling! Who laid the egg?"

Of the three, Curtis was by far the best athlete. He could outrun them and in activities requiring eye-hand coordination and quickness, he excelled. He loved to compete and continued to play softball into his forties. His natural competitiveness made him the only viable challenger to Jack's authority. I think he never fully acquiesced to his brother's dominance. Shortly before they joined Troop F I observed one of their wrestling matches in our bedroom out of sight of Mamma and Daddy. It was a draw as I recall.

Loyd, the inquisitive string bean, would learn to get by more on guile and charisma than on hard work and tenacity. He spent a lot of his early childhood in the kitchen with Mamma. He was fascinated by the writing on flour sacks and by words on other containers and the few books we had. He began asking questions and sounding out words and, by the age of four, was reading. His reading aptitude convinced Mamma and Daddy that he should be in school. At the age of five he entered first grade and within a few weeks and after his sixth birthday, he was promoted to second grade. He was also keen on arithmetic and would become an accomplished mathematician. This talent would help his military career, as much later he would command Nike missile units.

Loyd had a severe infection in his childhood that weakened him and he would become a frequent companion to Daddy on his peddling trips, leaving the older, stronger boys to do the work, much to their displeasure. This advantage didn't escape him and legend has it that he was able to parlay his condition into avoiding the sweatiest jobs for much longer than necessary.

Unfortunately, Loyd inherited the pugnacity of his Holder ancestors. This trait was even more prevalent in the Staggs branch of Mamma's family tree. His smart-aleck attitude enhanced his tendency to find trouble and resulted in numerous combative encounters as a young man. Jack was a great influence in his life and was able to keep him on a wobbly straight and narrow path.

Academically, the three were good students. Jack applied himself more than the other two. Curtis was intelligent enough, but his quest for a good time and mischief kept him from excelling in academics. Loyd, with his academic talents, was able to do very well in school and still have a good time. I doubt if he ever fully tapped his full potential, which in itself can become a psychological burden.

The Boys continued their adventure into the summer of 1941 with world events inexorably shaping the destiny of every person on earth in ways that they couldn't imagine. Jack and Loyd were in Louisiana learning more about becoming warriors and Curtis was in San Antonio learning to cook.

Jack's letters will tell the story of the war games. His first letter, August 15, from the Louisiana 'wild woods' spoke of the heat and dust.

"The chiggers are pretty bad down here. Some of the boys have a lot of them. There are a few ticks too. They just get on dogs and 3rd platooners tho.'

"We had an inspection this morning. It was sure a nasty one too. We all had on denoms [work clothes] & were wet with sweat, but the major inspecting us was nasty too, so it didn't make any difference. That Buzzard is just like a hog, he don't care how nasty he gets."

The condition of the horses and their adjustment to the new climate was a concern. They would ride distances of twenty to fifty miles at a time, crisscrossing the Texas/Louisiana state line from near Shreveport to an area between Jasper, Texas and De Ridder, Louisiana.

We received a post card from Jack dated August 22 that caused quite a stir. It read, "Well we are at the same place we were yesterday 7 miles south of Bayou. If you have about a dollar or two you could spare till payday I could use it. We get a chance to buy things along the road once in awhile & I cut myself short last month & ran out." No doubt that this shortage was caused by his loan business.

Daddy had been sick and out of work and had no money. Mamma had written and mailed a letter to Jack on the day this card came asking for a loan. I don't remember much of what was said, but I remember the stress it caused Mamma and Daddy. They had to get the post card back and find a way to get some money to Jack. Daddy went to the Garner post office and intercepted the card. I don't recall how they did it, but I am sure that they found a way to send Jack the money he requested. It could have come from a loan from Uncle H.D. because I recall a loan from him of eight dollars to buy groceries.

The men of the 124th Cavalry spent their time waiting in pleasant pastoral settings and playing war. Jack wrote August 29 from near Shreveport.

"We are still here about a mile west of Florine, Louisiana. We are supposed to be resting, but I haven't had time to wash a suit of clothes yet. I could just let my other work go, get some things done, but we had inspection today & I wanted to be ready for it. We made the worst showing that we ever did. We didn't have half of our equipment. We had lost it & tore it up & throwed it away & slipped it out till we just barely did have a saddle & horse. The Capt. really did get us told after the inspector left. We looked like a rebel army."

Jack mentioned that he and Loyd went to visit the artillery battery from Weatherford and saw several of their friends from Garner and Weatherford. Ben Hagman, Sr. was C.O. of the Weatherford artillery battery. Jack wrote September 2, "Well, we moved about 15 miles today south. We camped here about 2 weeks ago, close to Toro, La. It's a school house & store & church.'

"We left out this morning in a pouring rain. I didn't think we could roll our blanket and tear our tents down & get out in that mess but we just put our slickers on & went after it. All our saddle blankets & bed blankets got wet. We got here about four this afternoon & built up a bunch of fires & dried everything out. It quit raining about the time we left out this morning. We just took our time & stopped & fed the horses. This is more like a camp with a lot of fires going. This bunch is having a time. They are cooking that field rations coffee & eating what they have swiped from the kitchen. They are not hungry tho because we just got through eating supper."

Other letters tell the tale of their remaining two weeks in Louisiana and East Texas. He wrote September 7 from 'somewhere in Louisiana.'

"We just got through with a three day battle. It was against a regular army cavalry & some other outfits. My squad was in the middle of two hand to hand fights. You see-when there are no umpires right on hand you might not know who won & you have to fight to keep your guns. We had to once. One of my assistant gunners hit a 8th Cav. sgt with his fist three times then pulled out his pistol & drew it back & about 6 of them backed off. He is a mean Mexican. I think he is half Negro. He is black & looks like a Mexican & Negro but he can speak better Spanish than anything. A lieutenant tried to take a gun away from Len Bell, one of my gunners from Palo Pinto, & he jerked it loose & ran away with it. They had us outnumbered but there wasn't any umpire there & we wouldn't give up any of our guns or surrender. George Rankin said he'd like to feed our machine gun platoon raw meat about a month & take us into a real battle. We have got some rough boys. We went right down thru a thick woods in a gallop with Buzzard & Orazco (a Mexican), my other pack leader just skimming the bark off them pines. We really had some fun. Another time we slipped out to a road, dismounted & captured a whole platoon of red machine gunners & a 50 cal. gun. You asked about what we were. We are blues. I don't know how we came out as a whole but our troop won or came out even in every battle we went into the last few days. Boy we had tanks & heavy art. on both sides. We just met on a highway & spread out on both sides.'

"Well I guess you think I'm fight crazy but there isn't a thing to write about but that. Oh yes we left Bliss four weeks ago today & I think we have about three more weeks. I'll be glad when we get out of this hole & go back to civilization & some beds & tables & good bread, & shows, & showers & just a place to lay down & rest. That little Mexican village of Rio Grande City is going to look like heaven to us. Well we are about to go water horses. I'll finish this this afternoon."

September 11, Burr's Ferry, La.

"How is everything? We are making it o.k., I guess. Well since I wrote that first line we have gone a long way. We started from Florine the 10th & were at Burr Ferry bridge that night then we started up inside Texas & now we are at Himphill, Tex. It's sure good to be back in Texas. You can tell the difference at any time. When we crossed the river we got open pavement & ever since then we have found good bridges. We can trot right over these Texas bridges where we had to stop & walk over the ones in La. We made it from Weirgate here today & it's a long, hot way. We had forty even horses to fall out. We passed some out of the 112th that were dead as heck on the side of the road. It takes a rough old hoss to take it & not fall out. Mine just made it and that's all. Buz is riding a good bay mare & he came close to knocking her out today. He fell out with Mendoza to straighten his blanket & had to trot about 20 minutes to catch up & she stayed hot & winded all day. I hurt mine (He is an old 8th Cavalry horse) when we first came down here & he nearly falls out on every long ride but he never has.

He's a tough old goat. Buzz & I are pretty heavy on a horse that isn't a going Jessy.'

"We had an alert call last night. About one o'clock they came running up & woke up everybody & said that a tornado was coming in 30 minutes & we all got up and rolled our rolls & moved our saddles and stuff to high ground & lay down in our slickers & went to sleep. It never came of course but we were just getting practice. It hasn't rained for three or four days now."

Sept. 18, Center, Texas

"We are making it o.k. We are on the highway north of Center on outposts. We have our guns set up and a few riflemen out in front. Buzzard & a bunch of others are right in front of a café with two pretty good looking gals working in it. They stay in the place. That is F Troop headquarters.'

"We left our horses in Himphill Sunday & haven't seen them since. I hope we don't see them anymore till we leave La. The stable Sgt. & a few yard birds are tending them. Buzzard is writing some. He said he wrote this morn. I didn't know it or I would have waited till tomorrow. Did Loyd tell you? He won $15.00 on a punch board. He is pretty rich now."

Sept. 21, Newton, Texas

"We got back from our weeks experience as infantrymen. We spent one good restfull week without our horses. We rode trucks to Center and a lot of other little towns around there. We put up a defence around the towns and slept a week. We are back now tho & have ridden from Himphill down to Newton & are going on to Sanger tomorrow. I don't know if we will use the horses anymore or not. I hope not."

Sept. 27, Burr's Ferry Bridge, Louisiana.

"We left our horses again this battle & have been guarding this bridge all time. The Reds blew it up before we got here & Blues built a pontoon bridge across right above it. An engineering outfit built it at night & it is really a good one. Over 2,000 trucks came over in one day. They came across in a solid column from six in the morning till about five that evening. Sometimes one would stop on the steep grade on the Texas bank of the river & have to roll back. One old goofy boy let his roll back & like to have run in the river.'

"What do you want to do about the place? If you can get it & want it bad enough go ahead, but don't sell your hogs & cows. I'll let you have the money. I think that goon Mildred is trying to hold you up & make some money. She might have sent that bunch from up there to make you all take it quicker. I don't really think it's worth a thousand dollars. A real estate dealer could tell you. That's probably the reason they wouldn't loan but $700 on it. If it was fixed up with a good house & a barn tho it would be all right. It might be a good chance to get settled down & be where your Dad & mother lived & where you know everybody. That would be worth a lot. It would be a good place with a little work done on it. When we get out of the army we can help build a barn & fix the house. It will take a lot of work to fix it up but it would be a pleasure if you really wanted to do it. Go ahead & get it if you all want to & I'll let you have the money but don't sell your hogs & cows because you will need them if you want anything to eat & you know we've got to do that." He was referring to the Holder place where the family was now living. We didn't buy the place because of the price and the inability to get a loan. Daddy was resistant to using Jack's savings.

On October 3, 1941 Curtis wrote,

"Made Pfc. Today. I'll be making $36.00 a month. We are getting about 8 more cooks this morning. They made instructors out of 3 of us that have been here as long as I have. I have to take charge of the stoves and see that 6 of the new men learn how to cook on the stoves. A lot of planes and men came in today from Louisiana. The maneuvers are about over."

Oct. 7, 1941, Ft. Ringgold, Tex. letter from Jack,

"Well we are finally back. We haven't done much since we got back. We are cleaning our equipment up at ease. We aren't working like we would if we had Capt. Jack Dews here. We have a long time to do this & it seems reasonable enough to me to take our time.'

"The people here were glad to see us come back. You can emagine they would be tho. This post keeps the town going. The cafes have been going to town.'

"Some of the kids and men too have been drunk ever since we got paid. They are really having fun. Two of them are in here now. We have to keep them off the upstairs porch. They would fall off sure.'

..."Well I collected a little money this mo. I've just about quit loaning money. It doesn't pay.'

The next day Jack wrote again,

"I'm writing this in a café in town & there is music going on and a good looking Mexican gal sweeping under my chair so think nothing of it if I wrote 'I love you.' Or something like that. She is real sweet little white, blue-eyed half-breed. I think I'll marry her. No I wouldn't go quiet that far but she is good to hold hands with.'

"You all can put this money in the City National or use it if you need it. It doesn't make any difference. Mom if you & June or any of you need clothes or anything just use the money. I won't know the difference."

In his letter of Oct. 13, Jack talks about his money loaning business.

"Did you get my money? I'm going to send some more soon I think. Marlet Harrison owes me $35.00 & I think he's going to borrow the money at the bank & pay me. I'm going to see Lt. Hornsby tomorrow about loaning some money. I can loan $200 to these guys going home on leaves & get the Lt.'s guarantee on it & make $40 or $50 on it in about 15 days. That's not a bad deal if you ask me. If one or two get by me & don't pay I will still make a good profit & help a lot of the boys to go home. Any of them will be glad to pay $12 for $10. I used to charge 50% interest but that is unreasonable. I'd slug a guy that ask me to pay 50%." He was talking about monthly percentage rates, or perhaps even less time.

Jack wrote about a different type of news on November 2, "Curtis has already found him a blonde since we got back. They just moved into town not long ago." [He was referring to Joyce May, who becomes an integral part of the story at this point.]

Curtis wrote on November 10, "I finally got above 1st class private... I will draw $61.00 a month after the 18th. I'm getting $51.00 now. I will be T-5, same as corporal." He wrote again the next day.

"Playing basketball. I've lost 3 or 4 lbs. My teammates are Jack, Buzz, Bob Harrison, Sam Jennings, Hugh Warren, Mickey Crosland, and Fitzgerald. A family of white people live here named May. They have 2 or 3 girls about our age. Their Dad works in the oilfield. Their names are Lee, Mary and Joyce. I went with Lee, then double dated with Joyce and her boyfriend Jimmy Keller and I went with Mary. I'm going with Joyce about every day now. I saw her first at a carnival across the street from their house and a few days later she came out to F Troop barracks and I asked her for a date."

Jack wrote again on November 19 from Fort Ringgold.

"Well, it won't be long I hope. We got official notice today that we could get out of this thing pretty soon. The married men will be first. The ones with more than just the time we have been inducted & then we the ones with a year off. It will take four or five mos. for all of us to get out because only 15% per mo. is allowed to be discharged. I think we will make it by February or Mr. All of this you understand depends on wheather we keep out of the war or not."

U.S. destroyers had engaged German submarines by now. The troops were fantasizing.

By December, 1941 the troop had settled into a regular garrison routine. Jack wrote on the second that it was raining, a rare event at Ft. Ringgold. He indicated that he was doing well with his loan shark business, although he didn't call it that. He collected $255 besides his $58 pay. They were due a raise the next month of $6 each. He described a typical day in the barracks on a rainy day; cleaning rifles, playing poker, sleeping and writing letters. He admonished June, "You better learn to write English & write me some time. It would be o.k. if you wrote in Spanish but that Chinese just don't get it. Bill you write to us again too. You are a pretty good writer." He alternately teased and encouraged all of us. June was only five years old with no schooling.

He continually worried about Daddy's health and welfare. He inquired,

"Daddy, have you learned to drive the car pretty good. It is easier than that Model T after you get used to it. Have your teeth stopped hurting yet?"

Daddy's cheap false teeth bothered him for months until he patiently scraped away the troublesome pressure points with his pocketknife. Daddy had bought a Model A sedan with a gear shift stick which was a new wrinkle from the gear changing pedals of the Model T pickup that had been our transportation for years. I assume he had found a job, although I have no memory of it.

Daddy drove the old Model T so slowly that on the sandy roads of the time, R.C. and Herchel would jump off the tailgate where we rode with our feet hanging off and run along beside and could have passed us if they had wanted. It was the same vehicle that Herchel, R. C. and 'Son' Finger, a cousin, had taken to the woods while Mamma and Daddy had gone to town with Uncle H.D. They were supposed to be cutting and hauling firewood to the house. But they took advantage of this rare opportunity to drive the pickup around the woods, albeit a little too fast. Herchel didn't see the stump in the weeds that he hit. The force of the impact was enough to cause the vehicle to turn over, but without injury to the passengers. They were able to lift it upright, but it wouldn't start. I don't remember if they told Daddy about the wreck while H.D. got it started and I don't recall the incident having anything to do with trading it off.

The Model T was also involved in the Junior slicing incident. Before the boys joined Troop F, they were home on a Saturday. Mamma sent Loyd to Garner to get some Polly Pop and ice. Roy had to go with him. He begged Loyd to let him drive. Instead of stopping to trade seats, Loyd moved over while Roy climbed out on the running board to get to the driver's seat. Loyd was guiding the vehicle with his left hand just as the front tires hit an especially deep and loose bed of sand. The wheel cut sharply to the left and jerked the pickup into the ditch. By the time Loyd got the little truck under control it was running along a barbed wire fence slicing through the skin and muscle of Roy's rib cage. We heard his cries from the house. We ran the half-mile to the sight of the accident. Roy's side was laced with long and deep lacerations, but no broken bones. He carried these heavy scars through life and they were the source of some tall tales to the more gullible audiences about his adventurous exploits.

It would be five days until the Japanese changed the history of the world and alter the future of the Knight family profoundly and forever.

CHAPTER 7

PEARL HARBOR AND WAR

MOST F TROOPERS had taken advantage of Sunday off duty and had slept in. Curtis had been on the breakfast and lunch shift in the mess hall and had helped prepare breakfast for less than half the troop. By lunchtime most of the troopers were up and eating. Loyd and other troopers had been to town the night before and had spent a little too much time and money at Mexico Café. They were trying to shake the effects of the night's indulgencies. All movement seemed in slow motion and the usual banter was subdued. As they drifted sleepily into the afternoon all hell broke loose.

Their indolence was suddenly shattered by the excited voices of Mary and Joyce May. Some of the men walked out on the veranda and listened. They caught the words 'Japs" and 'Pearl Harbor'. In moments the message was clear.

"What the hell is Pearl Harbor?"

"Where the hell is Pearl Harbor?"

The men finished dressing and hurried to the day room where a radio could be found. As they listened to the static laced radio transmissions from Hawaii and the stateside commentary, the whole meaning of the message became very clear. The United States had been attacked and we had our war at last. Reaction from the troopers began to blend with the excited radio reporting.

"Those sneaky, slant-eyed bastards!"

"How did they manage to get to our guys without them knowin'?"

"I heard the guverment was about to get things settled with them."

"That was probly a smoke screen so they could sneak up on us."

"Hell, if those boys in Hiwaya had as gooda time as I had last night, they wadn't in no shape to be thinkin' about Japs."

"I God, I reckon somebody orta knowed!"

The simple questions and concerns were the first to come to mind by them and the nation. Their debate will last through history. None of the troopers could give a rational answer to these questions, but one thing was sure; any reservations about the U.S. becoming involved in a foreign war was instantly and wholly shattered. There would be no more speculation about the 124th Cavalry Regiment being de-activated or about individuals being sent home.

Corporal Jack L. Knight had quietly withdrawn from the maelstrom of the day room and sat quietly on the edge of his bunk, his mind rolling the new possibilities confronting him and his family. He was ready to do his part, but he worried about Curtis with his recently acquired serious girlfriend and Loyd's youthful exuberance and penchant for aggressive adventure. Most of all, he worried about Mamma's anxiety; how she could handle a war with so many sons, with her pacifist views.

Pfc. Loyd W. Knight approached with uncharacteristic solemnity. He could be a wise-cracking, belligerent smart-mouth, but he was highly intelligent and the radical change in their status demanded serious thought and he was in no mood for his usual pugnacious proclamations. This was not the same as picking a fight with alley rats in Mineral Wells or the E Troopers at the Mexico Café. He needed to get his nineteen year-old mind around this. Jack would help.

"Well, John, looks like all hell has broken loose with the Japs," he said as he sat on the next bunk, facing Jack.

"I don't see how we can stay out of it with the Japs. Those bastards have been asking for it a long time now. You know Churchill has had Roosevelt in his corner for a while now. We might have to fight the Germans, too. We're not ready for all that. We just don't have the army to do much and now we don't have the navy. Maybe we can put off fightin' the Germans for a year or two until we're ready, but what do I know. It won't take long to find out, though. Roosevelt and the military big shots will figger it out. I'm ready to do whatever they ask me to do." Jack mused.

He continued, "I think the Hundred Twenty-fourth is ready to fight now, but I guess we wouldn't amount to a fly speck on the wall compared to what the Krauts and Japs have been building the last five or ten years."

"Yeah, it's going to take us four or five years to catch up, I bet," added Loyd.

They didn't underestimate their own desire or ability to put up a good fight, nor did they underestimate the will and ingenuity of the American people. However, they, as well as our enemies, missed the mark on how long it would take for the country to respond and build the strongest military force in the history of man. The country was motivated by a just and holy war.

"Boy, I bet Mamma is beside herself. She worries about everything. Now she has something real to worry about," Loyd added.

Jack answered, "Yeah, her and Daddy and all their neighbors and everbody else will have plenty on their minds now."

A group of troopers came up the stairs. They had worked themselves into a lather.

"Boy, I hope they send us to Hawaya tomorra. I'd like to shoot the shit outta some of them cowardly Japs!"

"I bet F Troop could whup a whole regiment of those little weasels. I would enjoy sticking my bayonet up their asses and give it a good twist."

"Hey, Knight boys, are you ready to kill some Japs?"

Jack smiled and joined in, "Damn right! We'll mow 'em down in a few months and send 'em back to Japan with their tails between their legs."

Loyd added his two-cents worth, "I can't wait to get my hands on them. I could whip a whole squad with one hand tied behind me."

It was time to get stirred up. Thoughtful contemplation would become a personal matter to be exercised in the privacy of the mind. It was time now to get ready for war.

Mamma and Daddy and the five kids at home had returned from morning worship services at Holder's Chapel. We had dinner, and were entering the usual somnolence of Sunday afternoon. Daddy, still sitting at the kitchen table, turned on the recently purchased battery powered radio to catch the news. It was shocking. A hush fell over the kitchen as the full force of the day's news penetrated the soul of my parents. We drifted back to the table. All of us instinctively knew that it was no time for the light-hearted chatter or for the aggravation games we played with each other. Not much was said as we listened to the ongoing reports. But Mamma and Daddy talked quietly of war and its effect on The Boys as they finished cleaning up the kitchen. The four of us boys moved outside, as was our usual practice, and headed for the cribs by the cow pen.

The cribs were constructed by Pappy Holder, our maternal grandfather. They resembled small-scale log cabins without windows. Two of them were placed side by side with enough space between to drive a team-drawn wagon. A single roof covered both. Here we stored corn, sorghum, corn stalk tops and other cattle and horse feed. The harness for the draft animals was hung in the breezeway.

Out of earshot of Mamma and Daddy, Herchel and R.C., now seventeen and fifteen unleashed their quiet comprehensive vocabulary of cuss words directed at the Japanese. I could tell they were very perturbed because their cussing was intense and their choice of epithets came right off the top shelf of the cussing hierarchy. They invoked the wrath of God upon the Japanese and declared their questionable ancestry and legitimacy. Roy and I sneaked in some of our seminal repertoire of cuss words. But, being only ten and eight, we still couldn't reach the heights that the two older ones could achieve. Most of our inadequacy came from the fear that our newly found liberties might be squelched by the older brothers or that somehow Mamma's super-sensitive hearing might extend a hundred yards to our hiding place. We did, however, enjoy that bit of freedom if we didn't fully understand why we should have these opinions of the Japanese. Furthermore, protocol demanded our position in the hierarchy of cussin' had to be honored.

We were held in check also by our instinctive knowledge that somehow Mamma and God would know what we said. In our limited experience and theological insight we weren't convinced of the nature of God's punishment, but there was no such doubt when contemplating the form of Mamma's wrath. I knew what she meant when she said, "Billy Rowe, if you don't behave, I'm gonna land on you!" I never developed a taste for being 'landed on.'

Then followed the serious thoughts of the ramifications of the day's events. With Mamma's praying and Daddy's rationalizing we managed to get through the next few days and when no immediate disastrous events hit us we resumed our routines. In my childish optimism I developed the notion that my brothers were in some way immune to harm. I loved them too much. I would wish them through unscathed. Maybe they would even get through this war without having been the target of enemy bullets or having taken a life. I know Mamma was in constant prayer and Daddy occasionally. It appeared to be working. My brothers' decision to join the cavalry proved initially to be a safety net. Did my brothers have divine protection?

Jack's letter of December 8 reflected popular opinion and described the somewhat confused military response at Fort Ringgold.

… "Well it seems like we are at it, doesn't it? Well the sooner, the better. We have already lost valuable time. We can mow 'em down in six mos. & be back home.'

"We are going out on outposts tonight. I don't know where. We are going to stay a week & another bunch will go out. I'm not taking any writing paper so Buzzard will have to write. He isn't going. Curtis is going to cook.'

"Well, I've got to go. Don't worry about us because theres no danger. If you want to worry do it about some kids playing in the hollor or climbing trees. It's a lot more dangerous."

The Boys had been well trained to face unexpected adversity. Their character had been forged by a life on the farm, economic depression, seven siblings, college without money, a year of army training with those mule headed horses and the arbitrary rules. In December, 1941 the thought of war was a minor impediment in the road of life. They were tough and smart. They were athletic, disciplined and strong-willed. They had Mamma praying for them and they would prevail. Bring on the Japs and Krauts.

The status of the 124th Cavalry Regiment reflected the general conditions of the country and the national political scene at the time. We were confused, yet ready to do something, anything, toward winning the war. Jack's letter of December 11, 1941 is enlightening by the ambiguity it reflected.

… "We don't know where we are going, but everybody thinks we are going to Ft. Bliss. The captain said today that we were going to New Mexico to help guard a dam or some big electrical plant. I will be glad when we start. I think we will go to Ft. Brown first, and go on from there with the rest of the 124th Cav…'

"Some of the boys said that the Negros in Camp W[olters] were running off since war was declared. They ought to shoot 'em. Heck, this bunch don't care. They are ready to go. We should do some pretty good scuffling right now. Tell June that we are gonna scuffle with those Japs. Boy aren't those Russians pouring it on Germany?"

The story about the Negroes was only rumor, but reflected the false assumptions of most of the military brass and white folks in general that Negroes were not suitable for combat, absurd though it was.

By this time Germany had extended its army into an unmanageable territorial conquest with fronts in North Africa, the Balkans, France, Norway and Russia, and the day before Jack's letter, Hitler had declared war on the United States. Japan had moved far into China, Southeast Asia and the South Pacific. (WWII, American Heritage) With maniacs calling the shots, they were doomed to fight a losing cause for another three and a half years, bringing their nations to ruin; an astonishing indictment of nationalistic megalomania.

A few more days would allow the rumors to settle and reality to take over. The 124th was settling into a pattern that was to be their primary mission for the next thirty months; guarding and training. Jack describes their routine in his December 16, 1941 letter.

... "Well I'm on guard again. Buzzard just went off. We are on down here in town at the power plant. They have doubled the post guard and we have this guard extra. We have 23 men on guard here and in the post together. Well I guess we'll get a lot of this. We heard today that we might stay here a long time. This guard is o.k. We have a guardhouse to stay in and a radio & cots & lots of magazines to read when we are off posts. The pvts.have to walk 2 hrs. & are off 4. The cpl. & Sgt. take 2 hrs. on & off but we don't have to walk. We just sit in the guardhouse & keep awake.'

"That Murphy is the last straw. Buzzard is about as bad. They are giving the gals the rush around here now. Curtis goes with one all time & Loyd has got 3 different ones he goes with. He's got a real good one about lined up. She is a Mexican with kinda blond hair & big pretty eyes. She's really a purty booger. I'm still the same old woman hater of course.' [Jack liked the girls, but cherished his money more.]

"How is Toughy & Jew making it with their basketball? I don't guess we will play any more this year.'

"Mom I don't know what for you to do with all that junk we sent home. You all can use mine if any of the shirts will fit. Dad could wear our shirts & socks now. I've got 2 brown shirts I'm gonna send Dad." Now that the war had started, military personnel could no longer wear civies at any time.

"I hope you all like your Xmas presents. Me and Loyd just happened to be in the buying mood Sat. evening & we thought we'd send some things. He got a good little clock from Dot the other day." Dot was Loyd's girl back home.

"They have stopped just any & everybody from coming in the post now. Civilians have to have a pass from the Post Comm. Only 15% of the troop can go off to town at once now.'

"We took some shots today for lock jaw. We took them at 1:30 & we had to go by & wake that Pot Guts up on the way to the hospital. He hadn't got up all day. They just work every other day.'

"Boy Russia is really giving the Germans heck isn't it. That's all that saved us. If Germany had put the energy & men into an English invasion attempt & won Russia over by soft soaping them we would have been sunk. I guess we & China would have had to fight it out alone. I really believe Germany could have taken the British Isles if they had tried as hard as they have to take Russia.'

"How is Granny & Granpa? Is Laverne still working? I guess that Sam is still working a week & laying off a week." This was in reference to Daddy's parents and his two bachelor brothers who lived with their parents.

"I just went to town & ate two hamburgers & some pie. I brought one back in my pocket to eat tonight about midnight. None of this bunch knows about it. I just came up looking ignorant as usual. They think I'm innocent but I really ain't.'

"We had our first casualty about a week ago. Our machine gun Sgt. Owens married a Mexican gal. He is sunk. He is the first F Trooper that has married one. I guess she's o.k. but my gosh he could have found a German or Jap gal if he just had to get married.'

"We can't get but 5 miles from the Post, so that will eliminate that Mission & McAllen going. They caught an E Trooper in McAllen the other night. I don't know if he was running off or not but he is in the jug now.'...

"I don't know if you kids will like your presents or not, but we will send some more things Xmas. Bill I didn't think but I know you will all eat the candy & you won't have anything to keep. Well we'll send you something."

They sent me a large metal spring top that was activated by manual pressure on a twisted metal strip plunger. Christmas at the Knight household was Spartan. A few days before Christmas, we would search for a small juniper in the woods on our farm. We called these small conifers that grew prolifically in the cross-timbers region of North Central Texas 'cedars'. Then we would nail two one by four planks on the trunk for a stand, set it up in the corner of the living room that doubled for Mamma's, Daddy's and June's bedroom. The meager, well-worn red and green wooden ornaments would be placed on the tree along with two strands of frayed garland, one red, one green and a few strands of tinsel. It was absolutely beautiful. We knew it would attract Santy Clause's attention, even though expectations weren't that high. The year before Santy could only spare a small doll for June and a single-shot cap pistol for Roy and me to share. Daddy's ability to support Santa was hampered by illness and the lingering effects of the Great Depression. Even so, there remained antagonizing days of anticipation. Daddy's new job and 'The Boys' generosity improved our prospects. We got better toys and an apple, orange and a few pieces of hard candy in one of Mamma's old rayon stockings hanging by the fireplace. What delight!

We were at war and security, discipline, self-sacrifice and a fighting mentality were being heightened. Two years would pass before any of the Knight boys would see combat and then it wouldn't be the F Troopers, but Herchel who was a seventeen year old junior in high school at Christmas, 1941.

Daddy liked to brag on the older boys hoping that we younger brats would catch on. Mamma was always praying for them. If they were physically superior, morally impeccable and had God's and Mamma's protection, why wouldn't they be invincible.

The mission planning for the 124th was in flux. Guard duty seemed to be its destiny, but the details were continually changing. Jack's letter of December 23… "Well we don't think we will ever leave here now or not. I don't think we will. We have a new Post Commander & he is really rough. He is an 8th Cav. Colonel. He has a boy in A & M. This old guy has been in the Cav for 30 years."…

"We are starting on a new program soon. One troop will take all the guard duty for two weeks & the other two will stay in the post & train. We will have post guard, city power plant guard & one outpost at Roma & train for 1 mo. & then go back on guard.'

"What kind of work are you doing over there Dad? I asked in a letter a long time ago if you had got used to your teeth & driving the car. Have you?"

Daddy was hired to work at Camp Wolters and was assigned to the painting crew. With this job and later one at the 'Bomber Plant' in Fort Worth, he would have a steady job throughout the war and would be able to pay off debts accumulated during the Great Depression. He and the family would not have to depend upon farming for a living, but would continue to farm. This was our way of life and without it, we wouldn't be able to enhance our financial status and the four youngest boys would be deprived of the joys of farm labor.

With the replacement of the Model T pickup by the Model A Ford sedan, we thought we were up-town even though it was ten years old and we had to squeeze seven people into its limited space. The four brothers had to sit in the back seat. June sat between Mamma and Daddy in the front seat. This situation was appropriated by my older brothers to torment me and help me 'build character'. Herchel and R.C., the two oldest, sat on each side by the windows. Roy and I sat between them, but there wasn't enough room for both our butts to fit at the back of the seat, so he sat forward on the edge and could see out. Sitting down at the back of the seat, I could see only the treetops and sky. They would start describing scenes from the passing landscape of great beauty or grotesque freaks of nature. When I tried to struggle to the front of the seat to look out, they would hold me down with their elbows. I was never able to see these magnificent sights. My age, temper and gullibility combined to make this a very satisfying sport to them.

On December 30, 1941 Granpa Knight died. This was my first experience of memory in dealing with the death of someone so close. I remember the confusion and sadness of the graveside service and the surrounding tombstones. We would never again be able to visit him at his small produce stand on the corner of a block not far from the center of Mineral Wells and watch him shell peas as he ran the pods through the rollers of an old crank clothes wringer.

The bad war news continued as the Japanese army and navy expanded their conquest of Southeast Asia and the South Pacific. Our family moved from the Holder home place to an adjoining farm, the Howard Place, to the north on New Years Day, 1942. The house was about the same quality, with a little less room. The most attractive feature of this place was its location on the edge of the bluff overlooking the eastern shore of Lake Mineral Wells. There was one laborsaving feature to the new place. It had a water well within fifty yards of the house. At the other place we had to carry water in two barrels on a horse-drawn sled from across the farm, about half a mile. We had to draw the water from the forty foot depth of the well in a tubular well bucket with a hand released valve on the bottom. It would take fifty or sixty of these buckets to fill the two barrels. With Daddy's civil service job at Camp Wolters west of us across the lake, we settled in for a reasonably comfortable life of school, hunting, fishing, working the small farm and Daddy bringing home a steady paycheck.

The news from The Boys was also encouraging. Daddy had gotten home from work and read the letter from Jack that Mamma had walked a mile to get from the mailbox earlier in the day.

"It looks like the boys are doing o.k. You could expect those boys to find a way to go fishing and hunting if they had half a chance," Daddy said.

"Lord, I hope they can stay right where they are for awhile. I doubt that the calvry will be used to fight. Looks like they'd be perfect to guard the Rio Grande. I really doubt that will get dangerous. I can't see Germany or Japan coming through Mexico to get at us. You just keep praying that things won't change. Roy, I don't know how long I can stand this. If this war keeps up all of our boys could get in it. I just don't know… "

"Now Marthy, you know it won't last long enough for Junior and Billy Rowe to go; maybe Herchel and R.C. Most of 'em might be lucky enough to not to have to fight. You know that half the boys in service never have to fight."

"Oh, I know, but I just can't help worryin' about it. I'll go get supper ready while you and the boys milk and take care of the stock. We need to get in some firewood. Looks like we might have a norther come in tonight."

Herchel and R. C. went to the barn and cow lot with two milk buckets. They put out the cow feed in troughs at the head of the stalls. Both milked, kneeling on one knee at the right side of the cow, constantly alert for the swishing tail and the chance that one would lift a leg in response to pain from pressure on a cracked teat. Sometimes this happened in the winter and they had to be treated tenderly. This action could result in bodily injury or a spilled bucket of milk. When they finished, they let the calves take the rest of the milk and put out hay for them and the horses. They would then 'strip' the cows for the remainder of the milk. The cows would hold up milk for the calves and there was usually some left after the calves had their fill. They carried the milk to the house for

Mamma to process. Our milk processing consisted of straining the milk through thin cotton cloth into crock-pots. This removed the visible pieces of cow dung and hair. The milk was then placed in the icebox for cooling.

Daddy took Roy and me to the woodpile. The small logs had been gathered and cut from the oak woods on the eastern side of the farm. We used a two-handled crosscut saw and a double-bladed ax. A log would be placed in a notch on a larger log with the end extended the length needed for firewood. Daddy and Roy manned the saw. I rode the log to keep it from rolling with the saw and provide the resistance needed to make the cut. I watched the saw move methodically back and forth, the sawdust building in a pile at the sides of the base log. Several of these short pieces would then be split by Daddy and carried to the house by Roy and me, to be stacked by the sheet-metal heater in the living room/bed room. Sometimes Roy and I would try our hand at splitting the wood, with little success. This was another right-of-passage on the farm. Herchel and R.C. were strong enough and it was their job to gather the dead limbs that had fallen to the ground and cut the smaller trees. Those logs would be loaded on the horse drawn wagon for transport to the woodpile. A mix of dead and green wood was needed to make it easier to get the fire started and to help it burn longer.

During this time I aspired to become a contributing woodchopper. I challenged myself to cut a log in two. I chose a log about twelve inches in diameter. I whacked at it, chewing away at a spot about a foot wide. I rarely hit the same spot in succession, resulting in a shallow indenture that resembled the work of a disoriented beaver. Day after day my self-challenge became less compelling and finally I found other ways to fill my spare time. If that cut was ever completed, it was by one of the 'big three.'

At Ft. Ringgold, Troop F had added another segment to its mission. They were now training new recruits. Some of them were from Wyoming and Montana and had worked as ranch hands. They could ride, but not army style. As Jack put it, "They can't post worth a durn." When one of them was told that they might be on the border for the war, he cussed and said he didn't want to be stuck on the Mexican border. He wanted to fight. Jack wrote, "He doesn't know that you work six mos. like a mule, to get to fight six minutes."

For Jack, when it finally came, it was fifty months of training to fight fifteen minutes. Discipline and hard work were needed to win the war, no exceptions; not even for a bunch of patriotic country boys serving in an obsolete army unit at a nineteenth century cavalry post with no planned function in the war, except to pull guard duty and train new recruits.

As young men will do without fail, F Troopers found ways to entertain themselves in the midst of all the tedium and long hours of work, Jack describes some of their leisure activities in his letter of January 13, 1942 from Zapata, Texas.

"We are having some fun fishing and trapping. We have lines set out all up and down the river. We caught one channel cat that will weigh about three pounds. If we had some good bate we could catch lots of 'em. The coyotes like to have carried us off the first night we were out here & we borrowed a trap from this old guy here at the customs office that he had been trapping owls with & set it last night. We caught a medium sized possum the first night. He was as black as a skunk. We turned him loose." And later he wrote, "We are having a good fishing spree. I guess we have caught 50 or more lbs. since we have been here. Ray Warren caught a buffalo that would weigh about 10 or 12 pounds & we have caught about a dozen that would weigh 3 and four pounds. They are all channel cats. We had enough to feed about 35 men last night for supper & we caught six pretty good ones since… "

Curtis wrote on the same day and emphasized another aspect of this trip to Zapata.

"We are in Zapata patrolling the border. Buzzard is in Roma. Our kitchen is a tent. We have 2 field ranges and an ice box. We have plenty to eat and tents to sleep in. It's a little cold sleeping outside now, but I sleep in the kitchen tent and it's pretty warm all night. We brought our cots and 3 blankets, a comfort and mattress. We have machine guns and rifles posted all along the river bank. Our Lt. like to have been shot last night. He was inspecting the guns and guards abut 2:00 and the guard had a cold and couldn't make much racket when he said 'halt' and the officer didn't hear him and the guard didn't recognize him until the officer saw the guard pointing at him with his pistol and heard his hammer click when he pulled back. The officer yelled and told him not to shoot. We have orders to shoot first and ask questions later. It was Lt. Davis. I wear a pistol and we keep our rifles in reaching distance while we are asleep. It would be pretty hard for anybody to come across the river if we didn't want them to. I don't know why we have to do all this crap, because there isn't anybody crazy enough to try anything. I guess it's a good practice for us. My gal said she would write me every day, so I guess I can stand it a week. She came out to the camp yesterday and kissed me goodbye. She's a sweet little darlin'."

Later, in his January 26 letter, Jack went on to describe the work at Ft. Ringgold.

"We are really getting down to it. We have school all time and ride & fight little battles & do everything else. We have got a bunch of new officers. Some of them are o.k. I think one of them is going to take over the troop. He is six ft. seven inches & plays basketball with the best of them."

In the same letter Jack continues his teasing and advice with the following tongue-in-cheek paragraph.

"Bill you pick you out a good place in those rocks and remember the first thing to consider in selecting a position is a field of fire. The second thing is cover and concealment. In other words get a place where you can see them coming from any direction & they can't see you."

I was eight years old at the time. Unknown to him, we already were experienced in this element of military tactics. In one of our escapades into the woods, Herchel or R.C. had discovered a stash of cigarettes in a rock pile near the intersection of the railroad track to the south and the main gravel road to the east of the Holder Place. The summer before we moved, they decided to get some of those cigarettes. By stealth, we soon discovered they belonged to a neighbor boy living less than a quarter mile down the road toward Garner. Since they had reached the age that they would dare take a clandestine smoke, they wanted some of Rheudell's cigarettes.

They reasoned quite unreasonably that a kid his age had no business with that many cigarettes and that, since they had no conspiring uncle to provide them as Rheudell did, that they were entitled to share in his unlikely largesse. This conclusion resulted in an unarmed para-military operation that the army commandoes would envy.

Roy and I were posted as lookouts, ostensibly walking the rails for entertainment as we did occasionally. Any sighting of Rheudell was to prompt a Bob-white quail whistle. Thus established, R. C. and Herchel crawled on their bellies from our woods across the rutted, gravel country road, down the ditch to the rock pile. They carefully slithered around rocks to the cache. A quick removal of a couple of rocks exposed the cigarettes wrapped in cellophane bags. They stuffed their pockets with packs, replaced the rocks and quickly made their escape. Roy and I walked back to the road and casually joined the two victorious miscreants with their pathogenic booty.

To our knowledge, this theft was not recognized or was not acknowledged by Rheudell, since an alarm would have placed him and his cohort uncle in harm's way. The incident never caused any of us any guilt. Some fear perhaps, but not guilt. After all, the kid didn't need to smoke all those cigarettes anyway. Besides, those cigarettes were the real thing and added to Herchel's and R.C.'s status as wantabe men; much more mature than smoking grapevines, cedar bark or coffee rolled in paper sack, which was our usual smoking fare.

Another paragraph in the same letter gives us Jack's take on the progress of the war, "Well the news looked good today. The headlines in Santone Light says that the U.S. & Dutch hit 18 Jap ships & a lot of other good turns for us were mentioned. Maybe we are going to start a little damage doing of our own...Murphy is in town as usual. I think he is gonna marry that ole gal. I guess she's o.k."

All of Roy's and Martha's offspring were dedicated heterosexuals. You may be curious about the intimate details of our private romances, but let me tell you, you won't get it from me. One's sexual habits are his or her most private possessions. I will not probe, illuminate or criticize the personal elements of one's life, especially that of my brothers or sister. There are two reasons for this attitude; it's no one's business and I would never be able to look them in the eye if I started speculating. No, there are two more reasons this won't happen. I don't have the guts to ask and if I started to guess, my two remaining brothers, I am sure, could still whip my butt. I fear that it would certainly happen, because I wouldn't resist Curtis and Herchel, even if I could. Therefore, here comes the sanitized version because that is all I know.

Joyce was Curtis's choice. Their romance started in October, 1941 shortly after Curtis returned from cook school. Curtis, along with several F Troopers decided to break the monotony of their garrison life in the pre-war fall of 1941. Sports and barroom activities became routine and when a small carnival came to Rio Grande City they decided to explore its amusements and the prospects of meeting some local girls. Joyce told me her version of what happened next almost sixty years later. As they strolled through the carnival grounds, he spotted Joyce, a young lady of beautiful face and shape who was in the company of another F Trooper. His gazed fastened upon her and momentarily caught her eye and as Joyce described it, their eyes locked together. As it turned out, it was to be for the rest of her joyful life.

She didn't know him, but she wanted to. Joyce later had a date with Jimo Taylor, another F Trooper. As she waited for Jimo to arrive, she noticed that he was walking toward the house with another trooper, that young man whose eyes she had connected with at the carnival. He was coming to ask her for a date, not knowing Jimo already had a date with her. He could have been going to see one of Joyce's two older sisters. Her heart pounded. She couldn't go with Jimo if she felt this way about a stranger. As they came to the door she quickly responded, "I don't think I'm going out tonight." Disappointed, they left to return to their barracks. Joyce's mother sat nearby.

"Mamma, what's that other boy's name?"

She answered, "Curtis. Curtis Knight." He had already dated both of her older daughters.

"I've changed my mind. I want to go with Curtis."

There was nothing Mrs. May wouldn't do for her girls. She was also a woman of action. She sprang into action.

"Let's catch them!"

Mrs. May and Joyce ran down the street yelling, "Curtis! Curtis!"

The troopers stopped at this unexpected, yet welcome change of events. Joyce, a little out of breath walked up and without ceremony or apology called Curtis aside and whispered that she wanted to go with him. Subtlety was not a part of her game plan. She took charge to force events to her notion of how it should be. That did it for Curtis.

He decided to take the direct approach so he told Jimo that he was going with Joyce and figuratively told him to take a hike. They went to a movie and became spiritually inseparable for life.

Much has been made of Jack's frugality, but he wasn't the only one of the brothers to place value in a dollar. This is evident in Curtis' letter of January 20, 1942 from Zapata, … "I like to cook out here. You know in the army, when you want to do a thing, just tell the 1st Sgt. that you don't want to, and he will see that you do it. So I told Sgt. Whatley that I didn't much like it out here and he saw to it that I came out anyway. I am still drawing $61.00 a month and saving $25 a month."

Jack continued to advise those of us at home. In his February 1, 1942 letter,

"How are you kids liking school? Why don't you get some maps & study them about the war. About the Phillapines & Malay & all the East Indes & Australia. You can get maps out of newspapers, or at school. It's also interesting to watch the Russian advances. You really should study that. We are having an hour lecture on the international situation about 3 days each week. They are made by officers that really know a lot about it, & are good."

Jack's concern for our welfare and financial improvement is reflected in the relief shown in comments from his February 12 letter.

"Boy, Dad I'm glad to hear about your job. You ought to do all right on that. Go ahead & wear all my clothes you want. Will you get paid by the hr. or by a monthly salary. Write and tell us all about your job." He was understandably excited about Daddy being able to make some steady money. In another paragraph of the same letter he voices the beginning of frustration.

"We got a card from Lorene today. [A Holder cousin.] She wants us to hurry & win the war. I wish I knew what to do about it. Looks like it's going to be a slow procedure."

In his February 17 letter, "Curtis got a 14 page bible from Mrs. Broom the other day. She wrote in a small hand too. She wants us to win the war too. I still don't know how we can & sit here on our tails. They are not trying to send anything over there. They are probably waiting until they can send along sufficient air strength for protection. Well it shouldn't be very long. They are making ships & planes pretty fast now.'

"That was pleanty rough, losing Singapore, I mean. If the Dutch loose Java I still insist we'll be five or six years maybe longer whipping those lugs. If they just had us over there we could knock 'em off in short order, but we can't do it from here. I expect McArthur to be put out of it in another week or two. They will really crack down on him now since Singapore is out.'

"Believe it or not Curtis missed a night going down to see the May gal. I think he was sick the next day. Well she came out to see him the next day. She is a pretty good little gal with a good personality & fair looks, but I'll swear I couldn't stand Hedy Lamarr every night, or could I?"

Mrs. Broom was one of Mamma's church friends whose claim to fame, at least to the Knight boys, was her practice of cooking a dishpan sized banana pudding for our 'dinners on the ground' at church functions.

Jack wrote about Curtis and Joyce often. I was beginning to get a little insight into this man/woman mystery. We received a little indoctrination at school at this time of the year with the exchange of valentines. We exchanged these cards that always invited the other person to be one's valentine. I never had a complete understanding of why the other boys wanted me to be their valentine, nor them mine. I had an inkling of the propriety of a girl with those notions, but still with little understanding of the whole mystery of this connection. The valentines were tiny little cards of various shapes that cost about a penny and are still sold today, but at a cost well in excess of a penny.

Curtis was still a cook for Troop F. He saw Joyce every evening that he wasn't on duty. The attachment made it difficult for him to leave. As Mamma called it, they 'sparked' until late hours for a schoolgirl, with little discouragement from Mrs. May. She liked Curtis, too.

Curtis would pull his early morning shift in the kitchen and then walk the half-mile to the May home to make sure Joyce was up to make it to school on time. Imagine the thrill to a girl of seventeen to be awakened by her Knight in whites.

Their ache to be together prompted Joyce to act in uncharacteristically brazen violation of school rules. Curtis would borrow a friend's car and drive by the high school at noon break. Joyce would dash to the car with the principal calling for her to stop. She had to be with Curtis, so she was. Her usual sweet nature and studious habits kept her from serious trouble. No doubt, the psychology of the time made this brief escape with a soldier easier to accept, especially since she was always back before the bell rang.

In the same letter of February 17, Jack related the following story.

"Here's a good one. There's some planes that patrol the river here from Laredo to Brownsville. We heard that one of the guys had been landing over in Mexico & buying chickens & things for about half price & bringing them back across without paying the customs officer. Now when we see them come over someone will say, 'Yonder goes that damn chicken thief.'"

Jack carried on a continual commentary on the pace of the war. In his letter of March 1, 1942, he had this to say, "...Well looks like we are about to get started in the East Indies. I don't think the Japs can take Java. I hope not. They are about to get the Burma Road tho."

On March 9 he mentioned two of his recurring concerns.

"I don't guess I'll send any money home this month. I am loaning it all out. I'll have about $400 loaned out when I get it all loaned. I've got $300 at home. Boy I'll have some money some time. If you all need anything I can get just tell me. I'm gonna send you all & the kids some spending money."

On a Saturday early in March the F Troopers had completed their work for the day, showered and dressed in their Class A uniforms. Supper was over and Rio Grande City called.

"Hey Oscar, let's go down to the Mexico Café and get a cervazer," Loyd yelled across the barracks.

"O.K., Buzz. I need a cold one after this crappy day. But you've got to promise to leave the E Troopers alone."

"Shit! You know I never bother those square heads," Loyd answered with his tongue squarely in his cheek.

They hustled out the door and in a few minutes were leaving the post for the quarter mile walk down town. They made straight for the Mexico Café. There were a few locals finishing their supper. Soon they would be gone and the place would change character with the arrival of the boys from Second Squadron, 124th Cavalry.

Loyd and Oscar Martin were the first troopers to arrive. They sat down in a booth at the back and right away the friendly little waitress arrived to take their order. Loyd didn't think it proper to call her Emmer or Gert as he usually called other waitresses. In deference to her Mexican heritage he said, "Hey, Conchita, bring me a cold one." Oscar ordered and they began to nurse their drinks and wait for the inevitable.

It didn't take long for other cavalrymen to show up with a corresponding exponential increase in the level of noise. Good-natured ribbing between rival platoons in the same troop prevailed.

As they began to work on their third beer, Loyd and his buddy Oscar, being the first to arrive at this stage, began a not so subtle shift in the tone of their remarks. In a voice loud enough to be heard over the chatter, Loyd began, "Looks to me like there are too damn many E Troopers in here. What do you think, Oscar?"

"They ain't too many for me," he answered, slightly lower.

"Come on! You afraid of those assholes?"

"Naw, I ain't afraid. I just don't want trouble."

By the time the fourth beer arrived Oscar was beginning to see the logic in Loyd's view. There were just too many E Troopers in the place and they were making entirely too much noise.

"You know, Buzz, I think you're right. There are too many square heads in this place and I don't like their smart-assed attitude."

In a voice that would assure that all present could hear, Loyd let them know, "The damn square heads are making too much noise."

There is always on the opposite side someone with little enough inhibition to answer. Out of the crowd came the reply, "Hey Slim, what you gonna do about it?"

The bartender had witnessed this pattern of behavior develop on many occasions over the last several months and quickly called the post M.P.s.

Loyd and Oscar started walking toward the E Troop crowd. He didn't know who had made the remark and it mattered little to him. He was just after an E Trooper. He knew all of them who were regulars at their favorite watering hole.

As he arrived, he said, "You sumbitch, this is what I'm gonna do!"

He knew that in a brawl he had to land the first punch. He also knew how to punch. He had no intention of investigating the source of the remark. The one nearest would do. He gave him a quick jab with a severe snap of the wrist and the hapless trooper went sprawling across the room knocking over chairs and shoving tables. Oscar got another one.

There was sufficient alcohol content in all of them to repel any notion of good sense and caution. A couple of the E Troopers grabbed Loyd by the arms while others approached him from the front with hostile intent. He kicked one in the groin and twisted lose from his captors. As he was going for another, the M.P.s came through the front door. Oscar took off and disappeared, leaving Loyd alone with his enemies and the new adversaries with those arm bands that meant trouble.

"O.K., Knight, calm down. What the hell are you trying to do, wreck the place?"

With his usual disdain for authority, enhanced by his anesthetized brain waves, he replied, decibels rising, "What about these assholes? They started it."

"We don't care who started it. We just want it stopped. Come on, let's go outside and get some air and calm down."

They helped guide him outside, his resistance subsiding.

"We don't want to cause you any trouble. Just lighten up. Can you make it back to the post O.K.?"

"I don't want to go back now. I want to finish my beer."

"You've had enough for now."

At that time, Curtis and Joyce, out for a walk, came upon the scene.

Curtis inquired, "What's going on?"

The M.P. sergeant replied, "Loyd got into it with a bunch of E Troopers. He's had a little too much to drink."

"Murphy, I'm not drunk. I know what I'm doing."

"Then you ought to know enough to stay out of trouble," answered Curtis. To the M.P.s, "Let me have him. I'll take him back to the barracks."

"I don't want to go back. It's too early."

"You don't need any more trouble. Just shut up. You want to go to the guardhouse?"

Speaking to the M.P.s, Curtis asked, "Will you let me have him?

"If you take him and put him to bed."

"Buzz, come on. Let's go."

Unfortunately, the M.P. captain in charge of the Saturday night shift had heard about the fracas and fearing that it might get out of hand and wanting a break from the confines of his office, arrived. He walked up behind Loyd and placed his hand firmly on his shoulder.

"What's going…?"

Before he could finish, Loyd turned, thinking him one of his natural enemies and kicked at the same time. He noticed too late who was there. His momentum finished his original intent with the blow landing on the side of the calf of the captain, causing a slight stagger. That move automatically canceled the aforementioned mediation agreement between Curtis and the sergeant. Without hesitation, the captain ordered Loyd locked up.

Jack complains again in his March 28 letter and told of Loyd's scrape.

"It looks like we are stuck on this border. I'll bet we never get away from here. I don't know wheather I want to leave or stay. I'd like to be in on it when they start driving those Japs back to the North Pole. I don't want to do what we'd have to before we got to actually shoot a Jap. I guess that would be a pleasure that would overshadow all suffering. Well maybe they will come over here. We'll get our chance then, maybe so we go to them, huh?

... "Well I might as well tell you all. Buzz got on a couple of E Troopers or they got on him one. Anyway he ended up in the guardhouse. He won't stay long tho. He likes to fight too well. He always did. I think I'll wait till I see a Jap...Oh yeah, he won the fight all right. He knocked two of them on their buts. Oscar Martin was with him and got another one but he ran off when he saw the M.P.s coming. He will get off pretty soon. Our troop commander is mad about it. He said Buzz should have a medal for fighting E. Troopers."

Two days later General McArthur left Corregidor by P.T. boat for Australia.

With eight of us, naturally there were a variety of personalities. No greater contrast was evident than in our temperaments; from Loyd's and Roy's penchant for aggressiveness to R.C.'s and my usual passiveness. We could be provoked, but we didn't seek combat. A good example of my nature can best be illustrated by the Easter egg hunt story. I am not sure of the year, but it was while I was attending classes in the 'Dog House' at Garner School.

Easter for us was largely a religious day, with little secular emphasis. Mamma would boil a few eggs and we would color them with crayons while they were warm. The older boys would hide them in the yard. Under strict supervision we young one's would leisurely search for them and then share them with the brothers.

My first real experience with aggressive greed came on the last day of school before Easter break in that year of 1941 or 1942. Our teachers in grade school had gotten some high school students to hide the candy eggs on a large vacant lot across the street, west of the schoolhouse. I was about to participate in my first public Easter egg hunt. The teachers lined us in a rank along the ditch by the hunting grounds. Expecting the same procedure here as at home, as the signal was given to go, I started my stroll into this beautiful pastoral setting with its fresh grass, wild flowers and a few clumps of prickly pear and scattered mesquite. I was immediately left in the wake of a herd of swarming, discourteous, little savages grabbing every Easter egg they could find. I was momentarily shocked into immobility by this hoard of vandals. I never recovered. By the time I figured out that I needed to move with more spirit to acquire any bounty, it was too late. Most of the eggs had been snapped up. But for my good eyesight, I would have been literally left holding the bag. I managed to find three or four of the well hidden eggs. I was more shocked by the other kids behavior than I was disappointed in the results of my search. I think about that scene every Easter.

CHAPTER 8

IMPATIENCE IN WAR AND LOVE

THE JAPANESE CONTINUED their conquests through the spring and summer of 1942. The Allies, led by American grit, had a few minor victories and draws, but which were great morale boosters for a nation still on its heels.

On April 8 General Wainwright retreated from Bataan to Corregidor, ignoring an order from General McArthur to take an impossible stand. His men were too weak to fight. Many of them were captured. Colonel James Doolittle led the bombing raid on Tokyo on April 18, with little damage. He received the Medal of Honor for this daring deed. Although minor destruction was achieved, the raid accomplished its mission. He and his men demonstrated that we could strike the Japanese homeland. The people at home were emboldened by their sacrificial act.

Two weeks later the Japanese captured Guadalcanal about a thousand miles northeast of Australia. This conquest would lead, a few days later, to the historic Battle of The Coral Sea. The battle was not decisive for either side, but was momentous because it was the first naval battle in which ships from neither side saw each other. The aircraft carrier assumed its dominant role in naval strategy. The U.S.S. Lexington was damaged and was scuttled after the battle, becoming the first American carrier to be lost in the war. The Japanese withdrew their fleet to the Central Pacific.

On May 6, Gen. Wainwright surrendered his beleaguered troops. The infamous Bataan Death March ensued. Americans had forewarning from the Japanese behavior in China that cruelty was a Jap tactic, but were still caught by surprise by the savagery of this inhuman action.

General Joseph Stilwell ordered a retreat of U.S forces in Burma to Imphal, Assam in India. The German armies had been whipped decisively by the Soviets in the winter of 1941-42, failing in their attempt to occupy Moscow. However, they planned an offensive in the summer of 1942. The German attitude toward the Russian people led them to the same type of atrocities that were being committed by the Japanese. They were slaughtering whole villages of Russian civilians.

Admiral Yamamoto, indignant from the Doolittle raid and disappointed by the failure of achieving a decisive victory in the Coral Sea decided on a bold move to establish Japan's dominance of the Pacific. He gathered a force in excess of a hundred warships and supporting craft for the capture of Midway. About half way between Japan and the west coast of the United States, the small island would serve as a strategic base of operations for the control of the eastern Pacific.

The Battle of Midway occurred during the first week of June, 1942. With the help of the genius of our cryptology analysts and the shrewdness of Admiral Chester Nimitz, the Japanese operation was not entirely a secret and would not bring the expected results. With the bravery and good fortune of our sailors and pilots we would gain our first decisive victory and switch the momentum of the war. We would soon become the aggressor and Japan would be forced into fighting a delaying war in defense of its homeland. (WWII, American Heritage)

The Boys remained at Ft. Ringgold doing about the same thing as before. Jack was becoming even more frustrated. He wrote on April 7,

"I'll swear there just ain't nuthin' to write. This is a dead dump. I wish the Mexicans would attack the post. Maybe we would have some fun. There's no such thing tho. You can't tell one side of the river from the other. The people here don't even think of such a thing as fighting. We ride right down to the river bank & swim in it & fish & they are right across, not over a hundred yards away doing the same thing. They farm some pretty good fields right across the river from the post here."

A change of status for Curtis is related in his letter of April 7.

"Been out to hill 258 crawling around on our bellies like lizards. The darn thorns and cactus are so thick that we can't step without getting stuck. I guess we should be glad it's not bullets tho. like some of our men are dodging. I'm not cooking any more. I got to keep my rating. A guy that went to cooking school is taking my place. They wanted me with the troop on the line."

Loyd's experience in the guardhouse apparently taught him there were limits to his delinquent conduct. At age nineteen with eighteen months in the army, he decided to use his intelligence and guile to become a serious soldier. Jack bragged on him in his April 19 letter.

"I'm listening to the army hour. I don't know but I think it comes on every Sun. eve. We have been studying exactly the same things those machine gunners had, about the cone of fire & the beaten zone & all about firing. Buzzard is a gunner in the heavy M.G. Section. He really can learn anything quick. He knows more about those guns now than most of the corporals. I emagine if we had about three divisions like him in the Phillapines now the Japs would have a pain in the neck.'

"It was sure good the way our bombers gave it to the Japs in Tokyo, wasn't it? I wonder how they got there. They must have gone a long way."

Jack continued to work hard to be the best soldier he could be. His leadership qualities were not overlooked by his superiors. Trying to be modest about his accomplishments, he wrote on May 3, 1942.

"Well you will be addressing your letters to Sgt. Knight pretty soon. I am taking over the duties as a sgt. in the morning but I won't be officially sgt. till the papers come in. It will be a week or two. I want to wait about coming home till I can wear my three stripes. Isn't that goofy? Well you would understand what ratings mean if you were in the army. It means quite a bit to get another stripe, not to mention the extra money."

"Say I bought me a watch last night. It's a Bulova, 17 jewell. Cost $33.75. Boy it just played heck with a couple of my pretty 20's. I think I'll just wait & bring my money home. I did pretty good this mo. on collecting. I think I told you about us all buying a bond per mo. I really do like my watch. I needed one pretty bad too."

Mamma and Daddy were proud of The Boys. Daddy liked to brag about their accomplishments. Jack's success and frugality helped validate him as a parent. He hadn't accomplished a great deal in acquiring property or financial security, but he was very pleased that his boys could do it. I was taking all this in and I'm sure that it helped shape my life and in my mind for decades, the Bulova watch set the standard for time keeping. However, they had some concern for Loyd. There was reason for their worries, as indicated in Loyd's letter of May 11, 1942.

"Dear Folks,

"I got your letter the other day but have been too busy to ans. it. We sure have to work now. But you know I always was a hard worker.'

"I wish we could get to come home, but it looks like we won't now. One boy asked to get a furlough. He hadn't been home since he has been in the army and he hasn't gotten it yet.'

"I sure was proud to hear about Rowe being made a Sgt. I'll bet he will make a good one. I wish I was in his outfit. But I'm not even in Jacks. As far as me getting a rating, I could stay in the army forever and still be a pvt. I guess the reason for that is because I talk to the officers & non-coms in the same tone of voice that they talk to me.'

"Have you all saw Sammy or any of the McVays lately. I sure would like to see them and all the rest of that bunch. I guess I'll get to come back after the war is over.'

"I will have to quit now & rest up for tomorrow."

Write often,

Love

Loyd W.

In his letter of May 19, Curtis mentioned Joyce.

"My gal and her Mom said for me to tell Mama to come down here the next time she had a chance and stay with them. They live about 5 blocks from the camp. We could get a 3 or 4 day pass and be with them. They couldn't find a place in the world that they'd be more welcome than at Mrs. May's house. Bill

Ralph Aaron (Scroney) is going to marry Joyce's sister Mary the 1st of June. John thinks I'm fixing to get hitched too, but he's all wet...I haven't figured out how two can live as cheap as one yet." He would, out of desperation, very soon.

On the home front the family was routinely moving along with its activities. One of Daddy's favorite sports was fishing and his favored method was trot-lines. He would use shorter 'throw lines' and 'drop hooks' in the creeks. It was about this time that I was included in one of these family adventures. Being an avid sportsman, Daddy tried to blend the gathering of food and sport when he could. He hunted as well as fished with the goal always to bring as much to the table as possible. His leisure of choice during the war years was trotline fishing in Mineral Wells Lake.

About two hundred feet of heavy cord was used for the main line. The hooks were attached by smaller doubled cord about a foot long and placed about five feet apart. Much larger hooks at greater intervals were used in the deeper water. Smaller baits such as crawfish, small blue gills, sunfish we called perch, and minnows were used in shallow water to lure the smaller channel catfish and an occasional bass or crappie. Larger perch, crawdads and large chunks of rabbit meat and entrails were placed on the larger hooks. All this was designed to catch the large yellow catfish. Some of these weighed up to ninety pounds.

As I recall the events of this period, the sequencing becomes muddled and is surely flawed. My purpose is not to present a chronologically accurate picture, but to capture the state of mind and the essence of the family during those years.

I remember Curtis being with us. He can't remember this specific event, but I will record my memories. Uncle H.D. and his family were then living in Granny Holder's new house on the Holder farm. Daddy had a line out and baited. He, Uncle H. D. and Curtis decided to run the line. This had to be after dark because the lake was off-limits to public recreation of any kind since Camp Wolters was opened to train infantry for the military. An old wooden boat christened 'Mary Jane' was kept hidden in the cattails of a small inlet, out of view of the lake patrols.

We had to walk through Penitentiary Hollow to get to the lake. The hollow consisted of very large boulders broken away from the rim, forming crevices up to fifty feet deep. Natural caves could also be found. This is such a picturesque geological formation that it has become a Texas state park. Trails ran through and around these boulders and were difficult to negotiate at night without light. Daddy knew the trails and had keen eyesight. We had developed the skill to follow by sound and instinct. So we could follow without too much peril.

Daddy and the older boys always ran the line. However, on this night after much tearful pleading, Herbert, my cousin, and I were allowed to join in this adventure. As we approached the lake, stealth was S.O.P. All of us piled into the old, leaky boat. Daddy was in the front of the boat to pick up the line and remove the catch. Herbert and I then discovered the real reason for our presence. With the boat riding low in the water, it took on more water through the cracks in the dryer upper slats of the boat. Before we were well seated, Daddy said, "Bill, you and Herbert grab those tin cans and bail the water. Don't make any noise." This had to be done with coffee cans without bumping the boat timbers until the planks would swell and seal the cracks. In the dark, with little faith in that creaky, leaky old boat, I had to screw up all the grit that I possessed to step in. I planted my butt in the middle of the center seat with the notion that it would be safer.

As we approached the halfway point of the trotline, Daddy whispered to H.D. who was holding the line at the back of the boat. We didn't use nautical terms because we were farmers and I doubt that Daddy knew the difference between port and starboard. Herchel would soon fill those blanks in our vocabulary.

"H.D., we've got a big 'un on the line. Be careful. Let 'im play."

Daddy got on his knees with his hands under water as he carefully pulled the boat along the line and felt the tug of the big fish. The fish was on one of the last hooks and the weight at the end of the line had to be pulled from the mud and raised to get to the hook that held the fish. This gave the fish more freedom of movement. He hadn't been on the line very long because he had plenty of energy. He was pulling the boat around. The objective under these circumstances is to allow the fish to tug and swim until tired. If the fish pulls too hard, a practiced hand knows when to turn loose to allow more spring and stretch in the long line and to avoid having a hook jerked into a hand. After several minutes, he worked up to the tiring fish and gently the fish's head appeared above the water. That was the biggest fish any of us had ever seen. Daddy placed his hand on the fish's head to ease his hand into its mouth to clinch it and turn it over to render it less active. The head was at least a foot wide, indicating his enormous size. The fish then found another surge of energy and turned to the bottom. Daddy knew he had to turn loose and go back to the bank to pick up the line again. H. D. didn't have these skills or knowledge and he gripped the line in panic. He was strong. The fish was strong. The hook wasn't strong enough. We never saw that fish again, but we did see the straightened hook as the boat was paddled to tighten the line and relocate the weight.

There was the momentary silence of deep disappointment. Daddy didn't cuss out loud. Herbert and I were too young to hear the words appropriate for the occasion. He didn't reprimand H. D. It wasn't polite to fuss at company. I suspect that if Herbert and I hadn't been along and if it had been one of the older boys holding the line, his true feelings would have been vehemently proclaimed.

We didn't get the fish, but boy! We had a tale to tell about the one that got away. Daddy estimated that he would have been in the seventy to eighty pound class. The disappointment was from the sporting view, for the food value of the large, bottom-feeding fish was negligible. The meat was very fat and had a musky taste that was not as palatable as the channel catfish that were caught in the shallow, clearer water.

Several months later, as we prepared for another fishing binge, one of my favorite stories in the family folklore occurred. It was early June and the trotline fishing had started. Herchel had assumed the position of Daddy's straw boss over R.C., Junior and me in a successful effort to raise a variety of vegetables and fruit on a ten-acre plot. We had also assisted Mamma in preparing and canning beans, peas, corn, tomatoes, squash, potatoes, chow-chow and several varieties of cucumber pickles. We caught up and had a few days to free-lance.

We still had the old wooden rowboat hidden in the bushes and reeds by the lake. Daddy was ready to set the lines for a week or two of fishing. We needed bait. Earlier we had discovered a stock tank that our neighbors let us seine. It was filled with sunfish and bream, just the right size for catfish bait. Daddy told Herchel to take us to get bait that day and we would set out the lines that night.

We didn't have a car at home, so we carried the bucket and seine which was quarter-inch mesh, three feet wide and twenty-feet long. A small pole was tied at each end. It had floaters on the top and lead weights along the bottom. The older, stronger brothers would pull the seine through the water of the tank, a small, surface water reservoir. The little brothers' job was to follow the seine to keep it from snagging on whatever debris might litter the bottom of the tank. We would also lift the top to the surface if the tension forced the floats to submerge and allow the captured perch to escape.

We had the seine rolled around the poles and a couple of rusty five-gallon paint cans for transporting the catch. Herchel and R. C., being the oldest, carried the equipment. The tank was about a mile from home. The sandy lanes were hot, but not as hot as they became in August when the heat was unbearable, even for the tough soles we had developed on our bare feet. It would get even hotter in the afternoon. Every few yards we would have to walk in the grass with almost certainty that we would encounter grass burrs and goat-head sticker vines.

We made our way down the lane to the main road, which was surfaced with gravel. The road was followed about a half-mile until we reached the fence of the pasture in which the tank was located. Barbed wire fences can be tough to get through when you're the ages that Roy and I were. R.C. was tall enough to straddle the top wire. The fence was lax. Herchel simply placed one hand on top of a fence post and vaulted over. It was somewhat more difficult for Roy and me. We could lie down and roll under the bottom wire except for the grass burrs. We could step on the wires next to the posts, but that was a forbidden technique because of possible damage to the fence. We had to attack the fence directly, meaning a rare cooperative spirit was needed. One of us would get mid-way between posts, place a foot on the bottom wire and pull up on the next wire and allow the other to carefully project one leg through the opening, rolling his body slightly while pulling his other leg through, all the while trying to avoid butt contact with the barbs. We still had a hundred yards to walk across a fallow field to the tank. Many of the old worn-out farms had been abandoned because of the abundance of defense jobs available. We were ready for the thirty-minute job of seining the bait before the trip home. It was not to be.

Not only was the tank attractive to fishermen but to a variety of water snakes. As we approached the pond we could see their heads bobbing up all over the surface of the water. Herchel and R.C. wasted no time proclaiming the snakes to be of equal status to the Japanese. I understood the lambasting that the snakes received because I was there and could relate to the consequences that they had created. Roy, slightly lower in the cussin' hierarchy, exclaimed, "Shit fire and save matches!" I, being at the lowest rung in the hierarchy could only offer, "My Lord!"

It was agreed that we would toss out the bread and meal at the edge of the tank to attract the perch and dip from the bank to avoid confrontation with the snakes. On the first dip, we got two perch and a half-dozen snakes.

"The bastards have eaten the perch," was the consensus statement of Herchel and R.C. Roy was allowed to cuss occasionally and this qualified as one of those times and he took full advantage, exercising his prerogative to the maximum with, "We ought to kill them durn shitheads!"

I knew the words, but I had not reached the level of protocol that would allow me to show my manhood through the art form of adolescent cussing. I couldn't go beyond my initial contribution, so I remained silent, waiting for a solution.

Responding to Roy's heated proposal, R.C. questioned, "How we gonna do that, you ignert fart?"

Herchel had an answer. He quickly said to me, "Gopher, go get the twenty-two and all the shells."

I simply didn't argue with Herchel. I had the knots on my head to prove its futility. I took off. My mission was to go home, convince Mamma to let me take the .22 caliber rifle and a box of cartridges with me to the tank; no easy task. Mamma was also disappointed and with copious safety instructions, she relented and allowed me to complete my mission successfully. The others had managed to dip a dozen or so perch. Herchel, the boss both by age and strength got to use the rifle.

It isn't the easiest thing in the world to hit a snake's head bobbing in the water while standing without an armrest. He had some success, but by the time the shells were used only a half-dozen snakes had been killed. Too many remained. This challenge had ceased to be an effort to reclaim our perch hole. It was now a struggle of vengeance. Roy was then dispatched to the Garner store two miles to the east for another box of cartridges. All of us were throwing rocks and sticks, hoping to brain one of the snakes.

The next move by Herchel was a manifestation of his nickname. He picked up a club from a nearby trash pile and waded out waist-deep into the tank amongst the snakes. I was terrified. I knew that he would be bitten and would die right there. He wasn't bitten, perhaps because of the furor he created as he lashed out at every head as it popped up. We soon adjusted and R.C. joined him. They would addle or kill the snakes and pitch them on the bank. I would finish them off with my club if any life were left in them.

Roy returned with the cartridges and the snake slaughter continued. Shooting and whacking away, we soon made an impact on the numbers. We even found a den of them at the trash pile and bashed their brains out. We never once considered that the snakes might have a right or even a purpose for existence. They were snakes and they had fouled the best bait tank around. We were proud of our accomplishment. We wanted to show off.

We found some bailing wire and poked the wire through the snakes' heads or what remained of the heads. There were twenty-eight dead snakes ranging in length up to four feet. It took both Herchel and R.C. to carry them swinging from a discarded fence post resting on their shoulders. By the time we reached the entrance to our lane, they decided that they were too heavy to carry further. If we left them in the middle of the main road, we knew it would create some interest and show that the 'Knight boys' had struck again. There was no one else in the area that would do something like this. When Daddy arrived that afternoon from work, he knew his offspring had been at it and that it wasn't likely that he would be setting out a trotline that night.

During these years at the Howard Place, Mamma and Daddy would occasionally take June with them on their Saturday trips to town, leaving Herchel in charge. This arrangement inevitably led to attempts of mutiny. In this, I was a nonentity. R.C. was fairly even tempered and rarely challenged Herchel's authority. Under normal circumstances Roy had to acquiesce because of the six-year age difference and an infinite gap in combative skills. However, Roy had two characteristics that would at times erase those barriers and dissipate reason. He didn't take easily to hazing and he had a quick and volcanic temper.

Herchel, like the rest of us, loved to aggravate his brothers. During one of these 'free' Saturday's Herchel went beyond Roy's endurance and he retaliated with some cuss words that were taken a step too high on the hierarchy. He violated the sanctity of the cussin' protocol and they were directed toward this adolescent, temporary dictator. "Toughy' couldn't allow this breach of authority to go unpunished. Naturally, since the words came from Roy's mouth, he saw fit to stuff a wadded piece of paper in the same, while pinning him to the floor with knees holding down his arms.

Roy's brain completely ceased to function. Herchel, fearing nothing, released Roy. Roy immediately looked for an equalizer. He wasn't finished with this. The first thing he saw was a pair of large scissors laying on Mamma's sewing machine. Fortunately, Herchel saw this and decided flight was more prudent than fight and dashed into the kitchen just as the scissors struck the door facing and stuck. I witnessed all of this and I suppose my mind disconnected at the time of impact. I remember absolutely nothing of the incident nor about any consequences beyond that point. I suspect Mamma and Daddy never learned of this incident.

Jack had teased about his lack of a love life or any interest in the women around Rio Grande City. In his letter of May 17, 1942 he abandoned this line. He wrote, "I've got me a gal now, but I'm gonna quit going with her after today. I think she is the marrying kind. I don't go for that kind of baloney. What do you think about it June? Oh Yeah, she's a Mexican too… "

Jack continued his ongoing comments on the events and rumors of the war. On May 25 he wrote:

"Say, I guess you know that Mexico is about to get into the war. I don't know but it looks like to me that we are wasting our time staying here. We may go down in Mexico. Some Cav. from here went down there in the last war. I wish we could go down on the west coast about even with Mexico City. We may go to California. We may stay here."

Jack was a good soldier and became a great warrior but never heard of political correctness or racial sensitivity. Although couched as tongue-in-cheek, he could make some cruel statements. In his May 25 letter he added:

"June have you learned to read yet. You can write pretty good can't you. Bill can too but he won't write much. Who is your boyfriend now June Bug? If you come down here we'll get you a little Mexican. Boy these Mexicans are all right. I saw two little gals yesterday about the right size for Bill & Nig. They were just as brown & greasy as a pig. You know how brown I get in the summer. Well my gal was pleasantly surprised to find the other day that her hands were whiter than mine. She told her sister about it. She's pretty white….I'm like old Bob Harrison. I might have told you about it. He got him a new gal & Buzz ask him if she was a Mexican & he said no but she has been on the border so long that she speaks a little Spanish. She was a plain pepper bellie. Mine is too. Her name is L…. R….. She's a good gal & has a lot of brains. I'll send some pictures we made. I think I'll hunt me a blonde." Later, he would have a more serious relationship with another girl. Joyce and Curtis thought that this could have had long-term significance but for Jack's iron will not to get into marriage and leave a widow and orphan if he were killed in action.

The romances of the F Troopers were beginning to show results. Billy Ralph Aaron married Mary May, the older sister of Joyce. Joyce graduated from Rio Grande City High School in mid-May 1942 and moved with her family north to their hometown of Poteet, a small town twenty-five miles south of San Antonio. She and Curtis corresponded almost daily.

After a month of separation, Curtis couldn't take it any longer. He got leave and began to hitch hike home via Poteet. He was lucky. His third ride turned out to be someone in a truck on his way to Weatherford and was willing to stop over in Poteet. It was late that June night when they arrived at the May home in Poteet. Joyce was spending the night at her Aunt Virginia Davidson's house. They hurriedly got directions and drove there. Everyone had gone to bed. Cur-

Curtis & Joyce

Mamma & June visiting Rio Grande City, Texas

Curtis, Joyce, Mary, Billy Aaron
Glenda & Sandra present but not shown

Glenda

'Sgt.' Joyce

tis' loud and persistent knock brought Aunt Virginia to the door. She guessed that this brash young soldier was the Curtis of her niece's obsession.

"Hello, I'm Joyce's Aunt Virginia. Please come in… "

Without hesitation, Curtis said, "I want to see Joyce."

By now Joyce had heard the noise and voices and had hurriedly thrown on her clothes and appeared at the door. Curtis embraced her.

"Joyce, honey, I've got to talk to you. Let's take a walk."

They walked toward a nearby Methodist Church and sat on the front steps.

"Curtis, what are you doing here at this time of night? Is there something wrong?"

"Yeah, there's something wrong. I haven't seen you in a month. Joyce, sweetheart, I can't take this any longer. We have to be together all the time. You know we're going to get married some time. Why not now? Will you marry me now?"

The answer was a quick "Yes." She always allowed her heart to take control.

Curtis continued, "I'm going home to tell Mamma and Daddy that we're getting married. When can you be ready?"

"So you were pretty sure of yourself. What if I had said 'no'," Joyce teased.

"I'm a gambler, but I thought the odds were on my side," replied Curtis, grinning from ear-to-ear.

"It won't take but a few days. Where will it be?" Joyce responded, submitting to Curtis's directions.

"I won't have time to come back here for it. I don't want to make it a big thing. Why don't we do it at Rio Grande City. We can get Mary and Billy Ralph to be with us. We were the only ones with them. Is that o.k. with you? I'm sorry if I'm acting like a crazy nut, but I just had to come by and do something."

"Honey, of course it's o.k. with me. I'll be in Rio Grande City next week. We can get the license and find a preacher then."

They finished their conversation. Joyce returned to Aunt Virginia's for the night of head spinning thoughts and plans and little sleep. Curtis and the driver got a few hours of restless sleep and headed toward Weatherford.

Curtis arrived at our house in the mid-afternoon that day. We waited for Daddy to get home from work before his announcement. He told us that he was marrying Joyce May, the girl he and Jack had written so much about. As I recall, there was some anxiety from Curtis because he was the first to break the bonds of this tightly knit family unit. There was also a reassuring biblical response from Mamma that it was God's plan and the way it was supposed to be. After the tears was joy. All of us wanted Curtis to marry his pretty little Irish girl and bring her home. We embraced the notion fully and we already loved her like a sister.

Curtis thumbed it back to Fort Ringgold. Joyce traveled from Poteet to Rio Grande City on a Greyhound bus with her oldest sister, Lenora. Quickly, a license was obtained and arrangements were made. This turned out to be an ecumenical affair. Curtis proposed to his Church of Christ love on the steps of a Methodist Church. They found a Baptist preacher at a church on Billy Goat Hill in Rio Grande City to perform the ceremony. Did it last? Yes, it lasted until one month shy of fifty-nine years.

Jack's take on the situation is found in his June 20, 1942 letter in his normally straightforward way.

"Thought I'd write & tell you that that Pot Guted, barefooted Murphy got hitched. Buzzard & I went down to see him last night. They are staying in Caldwell's apartment. He has gone to cooks school in Santone & his wife went with him. Curtis & Joyce just sat & looked ignorant at us. We just teased 'em a little & went on to the show. They are as happy as if they had good sense. Oh yeah, I kissed her when I went in. Being as how he went & married her I'm gonna like her or know why. I'll expect the same from him when I marry this spick of mine.'

"She has gone to Brownsville now to see her cousin. I promised her I wouldn't go with anybody but blonds & brunetts & she said she wouldn't go with anybody but soldiers & sailors while she's gone so I guess we will make it o.k., because there's no red heads here anyway."

Jack described some experimental training in his June 25 letter.

"All the non-coms went down to Brownsville Tuesday & stayed till Wed. afternoon. Its 101 miles they say. It is a long way in the back of a 21/2 ton G.I. truck. We went to see a bunch of guys demonstrate how to float a lot of equipment & a Jeep & a Scout car on a canvas tarp. They put all of two peoples equipment saddles, guns & all on some pup tent, shelter halfs & floated them all over the place. They took a tarp off a 21/2 ton truck and ran a Jeep up on it & tied it up around the sides & shoved it in. It floated o.k. but boy it came close to sinking once. A boy was on top of it & if he hadn't jumped off it would have gone down. We have some pictures of all of it. I'll send some when they get back. They ran a scout car that weighs 9000 lbs upon a 20 by 40 ft. tarp & tied it up all around & shoved it in. It floated like a battle ship. I wouldn't have believed it but it can be done. This bunch saw some pictures in Life Magazine about it & thought they would try it too, I guess."

Jack gave a periodic update on their daily routine as he did in his letter of June 29, 1942

"We are not working too hard. Just about enough to keep in shape. We run a mile every morn. before breakfast, but even that is getting easy.'

"I'll tell you about a day of regular duty. We get up at 5:30 A to 5:45 and dress & make our beds. (I've told you this a dozen times.) We stand revilee at 6:00 & run that mile. We eat at 6:30. As soon as breakfast is over about 7:00, we put one half of the platoon to work on the barracks & latrine & the other half goes to the stables & cleans up the stables. We mop the floors & wipe off all lockers & sweep all the porches & police the area. At 8:00 we all ride out. We usually take all our machine gun packs. We ride in the brush almost all time now. They are making a golf course out of our drill field. (A national defense project I guess.) We ride till about 10:30 & come in. We clean saddles & pack harness till about 11:15 & then start grooming horses. We all have two wet with sweat horses to groom. We get through about 12:00 or 12:15 & it's always hot as heck. We go up and eat & are off usually about 15 to 30 minutes & fall out at 1:00. We have a nice little course of obstacles we go over almost every afternoon, about a half a dozen times. It has hurdles & crawls & cat walks & a wide ditch & a nine ft. wall. It's about a hundred yards long & you are pleanty worked out when you get over it. When we get about an hour of this obstacle course we go back to the barracks & have school on gas or weapons or some special work we're doing like practicing making those floats we saw those guys make down at Brown the other day. That's what we did this eve. At 4:30 we water horses & get ready for retreat. See we wear those old grease monkey coveralls for work & change to uniform, riding breeches, boots spurs & all for retreat. We eat supper as soon as retreat is over & are off after that if we are not on guard. So much for bull."

Curtis wrote on July 1 with an update on himself and Joyce.

"We moved out on the north side of town Sunday. It is a good big place and $3.50 cheaper. We will be living in the same house with Mary and Scroney."

Meanwhile, back at home at about this time, the littlest brother had learned to be mischievous. I had an epiphany the hard way; one that I would soon regret. I discovered on the same day within an hour time frame how easily I could lie in a convincing way and that there were consequences for doing it.

The Stubbs family lived a mile away and they had several children, all boys except twin girl toddlers. The oldest boy was Herchel's age. I remember Kenneth, the oldest one, because he would join the Navy about the same time that Herchel did. He would be killed when his ship was sunk at Leyte Gulf.

One of the boys was several months younger than me and, unbelievably, was more gullible. I contrived a plan to scare him with a clever lie. We had a female dog for a short time. She had died from a long forgotten cause and had been thrown off the rock cliff north of our house. Enough time had elapsed to expose her bones at the foot of a large boulder that had broken away from the rock cliff centuries before.

This young and trusting boy was at our house to play and be entertained by me. Not being one to plot, I was somehow given the inspiration to scare the kid, not because I didn't like him, for he was a good and gentle one. I didn't think through the cruelty involved.

"Hey, you want to see somethin' down in the holler?" I can't remember his name.

"I guess so. What is it?"

"It's the bones of a lady." I rationalized that this wasn't really a lie because we had named the dog Lady.

The kid was interested, but apprehensive. I believe that he didn't expect to find bones; that I was trying to trick him. With a little hesitation he followed me to the site. We had to get close to the edge of the cliff to peer down into the chasm. Both of us had to be brave to do this. As we looked down I pointed to the pile of bones twenty feet below that was Lady's remains.

"See! That's the lady's bones right there."

He didn't take time to examine the site. His belief was complete as he immediately burst into tears and headed home. I was surprised by his reaction and was reluctant to admit a lie. I was paralyzed by the reality of what I had done. My emotions were mixed. I was mildly proud that I successfully carried out the scheme and felt sorry for the kid at the same time. I did nothing but head for the house.

Mamma was in the yard and observed the bent and tragic figure heading down the lane toward his home.

"Billy Rowe, what's wrong with him? Why is he crying?"

Decision time. Could I lie to Mamma? I decided against that in a mili-second. I broke down and quickly admitted the whole sordid tale through tears of anguish. I hoped that this genuine posture would be enough to avoid a whipping. Immediately Mamma came up with a much more severe and humiliating punishment.

"Land sakes! Billy Rowe, what come across you? You go catch that boy and apologize and ask him to come back and eat dinner with us."

I took off. I had no other option. How do you apologize? What can be said? I caught him and fell into step. After many steps, I managed to utter, "That wasn't a lady. That was a dog's bones. Mamma wants you to eat dinner with us."

The boy never looked up and never stopped walking although his sobs diminished. In retrospect I suppose he was having an equal amount of anxiety in admitting to being scared and was perhaps ashamed of his timidity. This had to be enough. I turned and went back to Mamma.

"He's o.k. Mamma. He don't want to eat, though."

"Bill, don't you ever lie and don't you ever scare anybody like that again. Do you hear me!"

"Yes, Mamma. O.K., Mamma."

I wish that I could say that I have never lied since, but that would be a lie. I can say that every time I tell a lie, a small voice from the past whispers to me, "That's wrong. Don't do it." In my life I almost always seemed to get caught or found out when I committed a misdemeanor. I learned more that stuck from my mistakes. I should count myself lucky that way because I have made so many.

Jack wrote on July 4, 1942 about their trip to McAllen, Texas, a short distance down the valley from Fort Ringgold.

"Well we made it fine in the parade, and believe it or not we were the feature attraction. It started raining when we just started & got us all wet, but we didn't care. We don't pay any attention to the rain any more. We are staying right up in town tonight. There is a vacant block right back of the high school & we have our horses there. It's lot better place than we were last night. We will start back early in the morning.'

"I am at the U.S.O. hall the same place I wrote yesterday. They have a block of street roped off here next to the hall & we are going to have a big dance tonight. It will be the same thing exactly that you saw in M.W. that you wrote me about. It's good for the soldiers too, Mom, because if they didn't have this place the boys that like to dance would go to honky tonks & get into trouble. The girls they bring here are the best girls they can find. They have to pass before they are taken in. They don't just pick up anybody. They are all chaperoned by old women & go home with them. You may not think so but it really is the best thing I know of to let the soldiers have a good time & not have to be around bad influences. They serve cold drinks free too. There is about five or six big organizations gone together to make up this U.S.O. or whatever you call it. There's a boy over here now playing the piano. He can really paw it.'

"Well I kinda dred our ride back. I wish we would ride it all in one day. I don't want to stop over a night. We will camp tomorrow night where we did coming down here. It's at some irrigation lakes about ten miles west of Mission.'

"Buz has a date tonight with some little gal from Rio G. City. Pat Rounds & a bunch of gals & boys came down here to the parade today & will go back tonight.'

"Curtis & Billie's wives came over. They are going back in the morning. They have a room here about two blocks from the camp. I told them they ought to have made them stay home & Murphy said, 'My God, John, they like to get out of that damn Meskin den once in a while.' They are having as good a time as old Bill & N..... would over here. I guess they are. They are tickled to death. There are a heck of a lot of white people over here. Very unusual for this part of the state."

Later, on July 26, Jack was in a whimsical mood. He wrote, "There's a kid up here playing a French harp. It makes me think of the city place. I'll never forget those parties we used to have over there, when H. D. & Lockie & all of them used to play and Curtis & I would dance."

Those musicals were another type of party that could be tolerated by the more fundamentalist Christians like Mamma. There was no dancing by the adults, either square or round. The kids could cut up because there was no body touching between sexes. Of course, other folks were more liberal and there was dancing and clandestine drinking by the men in the dark, out-of-sight of the women and kids, or so they thought. At least, they sent the message to us that this activity was in some way wrong. This activity could occur at even the clean parties, if the Mammas were not vigilant. These events were still held during the forties.

At one of these musicals, the band took a break to go behind the barn to relieve themselves. One of them had a half pint of whiskey tucked into his boot. Each of them took a few sips, enough to break down inhibitions, but not enough to affect their music. Marvin McCracken was now the featured fiddler, but I never saw him take a drink. Lockey May had migrated to California for work in the shipyards. They came back to play a few requests. My favorites were the hoedowns 'Eighth of January', 'Under the Double Eagle' and 'Wagner.' Even old Red Walker and H.D. tried their hand at singing 'The Wabash Cannon Ball' and 'Hear the Lonesome Whistle Blow' or "When It's Peach Pickin' Time in Georgia' H. D. tried to emulate Jimmy Rogers. They did o.k., but nothing to compare to Roy Acuff's rendition from the Grand Old Opry in Nashville or the plaintive singing of Jimmy Rogers.

Red had gotten a little loose tongued after their break.

"I hang this git-tar around my neck so I can have something to do with my hands. Well, thar are other thangs I can do with my hands, but I'd embarrass you if I done it here in front of God and everbody. I say scratch where it itches, even if it's in your britches. Another shot of that Jack Daniels and I'd do it inny ways. Listen, ya'll don't tell Mamma I'm talkin' like 'is. She'd take a piss ellum switch to my butt…and I couldn't do nuthin' about it. Ennybody got a request. I thank I better start sangin'. Talkin' always gets me in trouble."

The band finished a set with my favorite hoedown, 'Under the Double Eagle.' This particular musical was being held in the home of one of the more liberal neighbors. 'Little Red' stretched and took the strap off his shoulders and carelessly held the guitar by the neck. He wasn't much of a guitar picker but he could sing.

"I God, I druther second to ol' Marvin playing 'Under the Double Eagle' than swim nekkid with Rita Hayworth," he said to the milling crowd as they moved outside.

Marvin answered him, "Red, you better feel thataway, 'cause you ain't never gonna git to go swimmin' with ol' Rita, nekkid or not."

"Hey, I know that. I'us just tryin' to say sumpin' good about ye. I didn't want to hug ye or kiss ye."

"Yeah! You'd have about as much of a chanch o' that as taking that swim with ol' Rita." H. D. was getting tired of this nonsense conversation.

"Hey, you guys ain't makin' any sense atall. We better git back in there and play some more while we can still wade through all this b.s."

With that they went back in and the adolescents resumed their activities. This was a good setting to practice their flirtations and steal a kiss. The more brazen girls would slip off in the dark with one of the adventurous and sophisticated boys to do some other mysterious things that I knew would not be approved by the Mammas. I was never brave enough to try to spy, but some of the older boys who didn't have older brothers with hard knuckles would sneak out to try to catch them doing something 'nasty'. They probably never saw anything, but they always reported that they did. From their descriptions, I wasn't sure at that time if I ever wanted to grow up to be like them. All of that nasty stuff sounded a little repugnant to me. I would learn better.

Mamma wouldn't allow herself to enjoy these risqué musicals, but she tolerated the milder ones. She had a few activities to get relief from the drudgery of farm work, housework, yard work, laundry work, cooking and taking care of the myriad needs of so many kids. She enjoyed church, especially the summer revivals. An activity common to most pioneering farm wives was quilting, a utilitarian hobby that was still practiced. We had our own quilting frames attached to the ceiling of the main room. Neighbor ladies would swap visits and spend their time stitching and gossiping. June and I, with the other kids, would play around the chairs and underneath the quilt stretched on the frame. It was just a place to play, but somehow the warmth and buzz of conversation made it a soul-enriching experience for both the ladies and the kids.

A new experience for the inland farm boys was described in Jack's letter of August 16 from Ft. Brown.

"We went with C Troop yesterday on a picnic & swimming party down on the beach. Boy that ocean is really big. We didn't see any submarines or sharks. I am really blistered today. That water is half salt. I don't like to swim in it, but sure is fun to ride those waves. They come in about as high as your head in water about 3 ft. deep. You have to jump up and keep your head up above them or dive into them and they will nearly knock you down. The sand along the water is as hard as a rock. We drove right along the edge in three of those big army trucks & they didn't hardly make a dent. The name of the place is Delmar Beach. It's about 2 miles above where the Rio Grande comes into the gulf."

CHAPTER 9

SALVATION, SIN AND LESSONS LEARNED

THE LAST HALF OF 1942 would see a gradual reversal of fortunes for the Allies. With encouragement from the victory at Midway, the industrial genius and energy of America began to strut its stuff. Troop and supply buildups in Australia and other Allied territory were gaining huge momentum. Japan wanted to thwart this activity. They began the construction of an airbase on Guadalcanal that would provide land-based bombers to attack this Allied lifeline.

Washington decided to advance its plan for an amphibious landing on the island. Parts of the First Marine Division went ashore to little resistance and soon captured the airstrip and named it Henderson Field. The taking was easy. The keeping was entirely another matter. The battle of wills raged on for six months. The marines were replaced by the army's Twenty-Fourth Division. The soldiers were as effective as the marines. Not only did our troops eliminate the Japanese entirely by February 9, 1943, but had gained vast and valuable lessons in jungle warfare and Japanese tactics. The first successful American land campaign and also the first land defeat for the Japanese left 24,000 dead Japs and 1,752 Americans dead.

Meanwhile, in the battle against Germany, the U. S. finally landed its first forces in French Morocco and Algeria on November 8, 1942. For some time failure was endured until Eisenhower could affect much needed training of troops and changes in command. General Omar Bradley with General George Patton and the British General Bernard Montgomery, himself a replacement commander, eventually whipped the Germans under Gen. Erwin Rommel and by late May, 1943, North Africa was cleared.

On January 22, 1943, Roosevelt and Churchill met in Casablanca for a conference in which they agreed to postpone the invasion of Europe because a cross-channel invasion wasn't feasible. This angered Joseph Stalin who desperately pushed for a second front in France to weaken German forces in Russia. They also agreed that unconditional surrender was the only option for Germany and Japan. (WWII, American Heritage)

At home we had reached the dog days of August in 1942. It was the time that we laid by the crops, meaning we quit working the row crops and waited for harvest. The gardens had declined and the watermelons and cantaloupes had seen their best days. At this time of year, one of our favorite treats was homemade ice cream. Daddy would go or send one of the boys to the store for a twenty-five pound block of ice to be wrapped in a thick discarded quilt for insulation. With an abundant supply of milk and eggs always available, Mamma would prepare the mixture. Well-ripened peaches were almost always available to add to the mixture. If not, bananas or just plain vanilla, were the choices of flavor. The ice would be chipped into several smaller chunks and placed in a tow sack. The larger boys would use the flat side of an ax blade to pulverize the ice. The six-quart ice cream bucket would be placed in the wooden freezer and the handle-turning mechanism fastened. The ice and salt mixture was then added. The quilt would be folded and placed on top with Roy or me on top of that. I thought my job was to hold the quilt in place to keep the cold in the freezer. I discovered years later that my weight helped force the worn gears of the turning mechanism to the worn gears on the bucket lid into closer proximity to prevent gear stripping and allow the bucket to turn. After I had children and with a new ice cream freezer, my boys still got to sit on top of the freezer. Somehow the process wasn't complete until a kid sat on the freezer and I'm sure that in some mystical way this enhanced the texture and flavor of the confection.

It was time for another summer activity to begin. The evangelical churches would begin the 'revival' season. We would always attend almost every night of the two-week revival at Holder's Chapel. We would also attend a few nights of the other church revivals. Mamma especially liked the Pentecostal revivals. Somehow, their more spirited services served her needs more than those of the more formal services of the Methodists, Baptists and The Church of Christ. She became a member of the Assembly of God along with June some time after the war was over. Daddy would later follow her.

Revivals were held annually for the purpose of spiritually uplifting the confessed Christians of all evangelical protestant churches and to bring others into the fold. The pastors would invite the more charismatic preachers to conduct the revivals. The services were attended by the whole family with an occasional exception for the older kids nearing adulthood. If they weren't saved by then, they were usually cut loose to make the decision to attend. Revival hymns such as 'The Old Rugged Cross', 'Jesus Saves', 'Nothing But the Blood', and 'Amazing Grace' were sung with great gusto at the beginning of the services accompanied by someone on the piano. This was followed by testimonials from the righteous and self-righteous. Somehow, it seemed to me that the ones who did the most talking were the ones who needed the most help. The sermon would follow. A minister wasn't much of a revival preacher unless he could evoke vivid images of hell-fire and brimstone for the unrepentant and the backslider. If one accepted Jesus Christ as one's personal savior and confessed one's sins one received salvation and was guaranteed a life everlasting in the presence of the Lord in heaven. Otherwise you were destined to spend eternity in the fires of hell stoked by Satan himself. In a Texas August one could readily relate to the

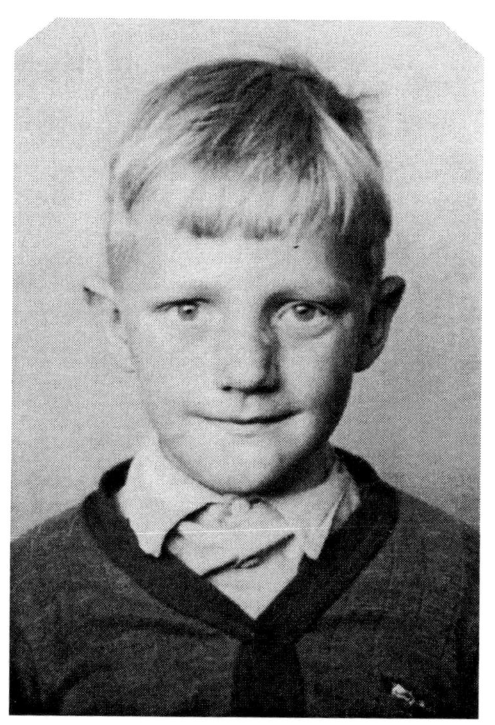

Bill, 1942

Bill in a pout and Ted

Bruno

Mamma and Ted

June & Ted the puppy

descriptions of hell. Our light was provided by Coleman lanterns that burned intensely, throwing out heat that only added to the discomfort. The only relief came from an occasional breeze drifting through the unwalled tabernacle and the energetic use of the paper fans with the wooden handles provided by the Baum and Carlock Funeral Home.

Preacher Hill, a kind old gentleman, who had been a Methodist preacher his entire adult life was invited to preach the revival in the summer of 1942. He was the father of Aunt Madell, Uncle Bill Holder's wife. I was eight years old and had been listening to the sermons, Sunday school lessons and Mamma and Daddy for several years. I had acquired some rudiments of understanding of the Christian faith and had acquired several layers of guilt for my naughty thoughts and deeds and more than a little interest in immortality.

I was introduced to Christian literature at Holder's Chapel. Miss Katy Scott taught the little kids class. We met in the corner of the sanctuary behind a curtain. We could hear the muffled sounds of the other classes. She would pass out small cards with illustrations of biblical events on one side and a short message and scripture on the backside. I loved those pictures because they were in color and everyone in them was so serene. I listened to Miss Scott and occasionally got the point of her lesson. We got to color some too. As you can expect, I remember some of the pictures, but no specifics of the messages. If I retained anything spiritually it would be merely a reinforcement of Mamma's daily lessons of doing right.

Perhaps Preacher Hill had mellowed with age or simply acknowledged that scaring the devil out of people usually led to only a temporary condition of righteousness. He didn't follow the usual pattern of loud and fearful predictions for my eight-year-old sinful soul. Instead, he emphasized the love of God through Jesus Christ. I don't recall the sermon, but I recall the gist of the summation,

"So you see that God is love. His son Jesus Christ loves you. You've known that since your first Sunday school lesson. Praise God! Now if you want to join in this feast of love or if you want to renew your commitment and strengthen your love for our Lord Jesus Christ, come to the altar. As Katy leads us in number fifty-three, come on down. 'Just as I am, without one plea, but that thy blood was shed for me, and that Thou bidd'st me come to Thee, O Lamb of God, I come! I come!' There is only one thing stronger that a mother's love and that's Jesus' love."

Katy Scott, in her version of alto loud enough for all to hear began the song. I didn't know anyone could love me more than Mamma. I knew Mamma loved His church and I knew that she knew a whole lot about Him. Maybe that's why she loved us kids so much. I was pulled down that aisle by a primal force stronger than me, more loving than man, to connect with Mamma's and Daddy's God.

One at a time Roy, June and I went to the altar. I doubt that I would have had the courage without Roy, and I was comforted by six-year old June and emboldened by Mamma shepherding us down the aisle. With the guidance and prodding of Mamma and Preacher Hill, accompanied by a flood of tears, we collectively confessed our sins and accepted Christ as our personal savior and joined the Holder's Chapel Methodist Church.

I thought my biggest problems had been solved. I would no longer have those naughty thoughts and would no longer displease Mamma and Daddy and the preachers and I would be reserved a place in heaven when I died. I just didn't want to take a swim in that lake of fire and I sure didn't want to be bossed around by the devil. Just remember that I was only eight and had no personal benchmark for these kinds of things.

For the most part I have been faithful to this commitment with considerable remorseful backsliding along the way. During those first few years I worried about every transgression from cussing to the dirty thoughts and deeds common to an eight year old, much of which had to do with the ugly finger. I didn't know about the powerful force of this appendage at the time. Real cussin' was not allowed around our house. Even imitation cussin' wasn't tolerated by Mamma. "Aw, shoot" was about the limit. 'Heck' and 'dern' might get by Daddy if it came from the older boys. We learned to cuss from each other and at school, with an occasional lesson from Daddy when he became angry, impatient or smashed a thumb with a hammer.

I have mentioned a hierarchy of cussin'. There was also a cussin' protocol. The younger kids were permitted to hear the older boys say the bad cuss words. Girls were never included. We knew not to say them around the older boys, but occasionally were brave enough to try one out on a contemporary. Of course, this was accompanied by considerable risk. When you are a kid you never know when the other kids will try to curry favor from a teacher or parent by telling on you. It took a very courageous or foolhardy kid to cuss around the sissy in the class. He always told on you. We had to reach adolescence when the peer group code was iron clad that we felt safe enough to really cut loose with the upper level vulgarities in the hierarchy.

I never heard Daddy say anything like the 'f' word or 'g.d.' I might have heard a 'damn' or a 'hell' or 'shit' from him after I was in high school. I suppose the rationale for that was that by then our eardrums had acquired some kind of discriminating filter. We were all grown before I heard Mamma say anything that could be considered naughty; and that was a whispered 'shit' after burning her hand on a hot pan.

Poverty helped accelerate the initiation of the boys' universal sin. We just didn't have much else to play with except sticks, stones and body parts. My memories include the first awareness that anything to do with this part of the anatomy was to provoke caution. To understand the full impact of this discussion you must understand that in the matter of laying on guilt, a southern protestant evangelical mother takes no back seat to a Jewish or Catholic mother. Very early in my life I was introduced to this emotional concept by pure coincidence.

We always had a couple of dogs of questionable lineage around the house. They served as watchdogs, copperhead snake alerts and squirrel hunters. They were always hungry and hung around the back door while at the house. They were more concerned with their own survival than with our security. Throw left over biscuits out the back door and each dog would catch one in his mouth before it hit the ground.

With the lack of inhibitions of a three year old, coupled with the absence of indoor plumbing, I chose to relieve myself in the yard next to our house instead of walking fifty yards to the outhouse. At that moment, our newly acquired half-grown Collie pup, Ted, happened upon the scene. Following the dog propensity to sniff a crotch and with ardent affection, Ted engaged in the activity with my full cooperation. It is no exaggeration to say that he had his way with me. This attracted the attention of my brother Roy and my sister June. An alarm went up with Roy yelling. You see, he already knew about the taboo. June added to the noise with her year old jabbering. Mamma appeared, agitated at the interruption and the unintelligible shouting, perhaps thinking that one of us had been maimed. What caught her eye was even more upsetting. With no idea that I was causing the disturbance I had taken no evasive action.

"Billy Rowe! What are you doing? I'll swan, what's gonna come of you!"

Silence. I didn't know what I was doing, but I did know what Ted was doing and to be honest, it wasn't altogether unpleasant or upsetting to me. I couldn't answer the question because I had no words to use or frame of reference for coping with this sudden shift in my status and Mamma's demeanor terrified me.

"Don't you ever let me catch you doing that again! If I do I'll land on you good. Now, you get in the house and behave yourself. Shame on you!" She saw no reason not to add an emphatic punctuation to her diatribe.

I was confused by this tirade as well as shaken. I knew she wasn't angry because I had peed in the yard. We did that all the time. There was only one other answer and that had to do with Ted's curiosity and the object of his affection. Somehow I guessed correctly that my sin was standing there, not moving away and allowing Ted to satisfy his olfactory senses. Two years later Mamma had her reason to land.

We had moved twice in that span of time, from the Crosthwaite Place to the Rawling's Place on the creek above Lake Mineral Wells to the old home place of Mamma's family, the Holder's Place. Fate would bring me together with Bruno, our latest dog acquisition, in a somewhat accidental role reversal. I discovered and became quite fascinated with Bruno's lipstick that appeared every time I played with him and stroked his head or stomach. My irresistible urge to investigate this phenomenon in a much more intimate and scientific fashion brought an even more cooperative spirit from Bruno.

Alas! I was found out again. This time by a little sister who had acquired language in the intervening years. June observed me during one of my investigations of Bruno's anatomy and response. No doubt she misjudged my activity and misinterpreted Bruno's gyrations as a reaction to pain being inflicted by me. She ran into the house shouting, "Mamma! Mamma! Bill's pinching Bruno's poochie!

Mamma knew immediately what I was doing and again overreacted. She landed.

"Goodness gracious alive! You're going to be the end of me! You stop that this minute. I don't know what I'm going to do with you." It didn't take her long to decide. She chose whatever weapon most handy to deal with such transgressions. This time it was the palm of her hand as she blistered my little behind.

"Stinkin' on it, Billy Rowe! You better not ever let me see you doin' that again! Ever time you do such a thing, I'll tend to you and I don't mean maybe. Good night nurse, I never saw such a thing." Mamma, usually reticent, could become aggravated to animation with violations of her Puritan code of conduct. She would use her entire repertoire of threats and admonitions.

With a load of pain and guilt, I dissolved into a pathetic heap of snot and tears, my innocence forever shattered and my vigilance forever heightened against the discovery of anything to do with that thing down there, whether on man or beast. My primal intuition was correct and was thence cultivated to a fine art. A year or so later a cousin educated me on the rest of the story and thus began a lifelong quest to regulate the demands of that tyrant, that disobedient, villainous appendage, that joyful wand of procreation and intimacy. So far I have been able to keep the mischievous miscreant under house arrest and to stay within man's law, if not God's. At least, I haven't started a war or caused a scandal.

I'm not implying that all guilt is bad. It can be a good thing as a warning signal against anticipated behavior, a safety net when tempted. It can be bad if it is laid on too thick, becoming oppressive, is never lifted and becomes a hammer. Some religion mothers use guilt as a weapon. I use the term 'religion mothers' not 'religious mothers' because I see a difference. Religion mothers aren't necessarily religious. They'll beat you up with the guilt. You may not be doing anything evil, or even wayward, but when a mean streak hits, duck.

Religious mothers lay on the guilt in a sincere effort to get you to recognize the difference between right and wrong. In most cases it works. One's conscience is enhanced this way. You always feel it when the signal goes off, especially if it has to do with the use of the 'thing'. But I'm afraid our response many times is akin to our reaction to the traffic warning light at an intersection. When we see it, we hit the accelerator to try to beat it, in effect ignoring its purpose.

The sweet thing about Mamma was that she understood creating guilt to be training for future decision-making. She loved her babies and always forgave. Henceforth, I gave her few occasions to forgive. I became ever more deceptive and vigilant in my erotic practices and avoided detection. I wonder if she thought I had become a saint or merely a more evasive sinner. What she didn't know couldn't hurt her, or so I thought. I learned later that she actually worried more about things she didn't know than those she did. She had more disturbing events to worry about anyway, those that embraced the world in man's propensity for war.

Before I go on with the war stories I will tell about a conversation soon after I joined the church in 1942. Mamma , when it came to interpreting the Bible, was a strict constructionist. She had a few interesting inconsistencies that she ignored. For example, she found lipstick a sin, but face powder was acceptable. Later she didn't apply the same condemnation to television as she did the picture show. Some smart aleck may even call her a Pharisee.

I had heard preachers blast us with threats of hell and entice us with the glories of heaven. Not a few times I heard one of my brothers and friends invite another one to 'go to hell.' The manner of speech and the circumstances gave me the impression that this journey was not a vacation. The conversation went something like this. In an effort to improve my religious education, I asked, "Mamma, is it o.k. to go to hell?"

"Billy Rowe! What do you mean?"

I was careful not to betray my brothers, because I suspected that it would cause stress within the household and ultimately result in regrets on my part. So, I prudently chose the preacher route.

"Well, the preachers are always talking about hell and fire. What do they mean?"

"Bill, the Bible says that hell is the place sinners go when they die if they wasn't saved first. People who go there spend eternity in a fiery pit, a lake of fire, without water and they gnash their teeth. They're right there with the old Scratch forever."

I breathed a sigh of relief because I wasn't a sinner anymore. I had taken that trip to the altar and was saved. I would find out much later about 'falling from grace'. I recall some heated debate between Daddy and Uncle Paul, a Baptist. It had something to do with the Baptists' theology of 'once saved, always saved' and the Methodist position that one could fall from grace. Of course, at the time I didn't understand clearly what they were talking about. In truth, I still can't claim complete insight.

Mamma didn't wait for the expected question about heaven. If you talked about one, you had to talk about the other. She didn't know about metaphors. They didn't teach that at Red Mud School where Mamma completed her formal education before the eighth grade.

"When we go to heaven, everything is real nice. There is no more sorrow and sadness. We'll get to live forever without sinnin' and we'll get to listen to the angels' choir and the Lord. We'll get to worship the Lord anytime, because we won't have to work in the fields or take care of the stock and you won't have to dry the dishes or churn the butter. We will live in mansions on streets paved with gold."

"Mamma, does the roof leak?"

"I don't know. The bible doesn't say. I don't even know if it rains up there. I always thought about heaven with white clouds."

Boy! This sounded good to me. What a payoff for that short walk down the aisle. I sure was glad Daddy and Mamma made me go to that revival at Holder's Chapel with that good old man, Preacher Hill, doing the preaching. I was a very lucky kid.

I have spent a lifetime hearing preaching and reading on this topic of heaven and hell. I haven't sorted all of it out, even yet, but some of what I was told when I was eight just doesn't seem to matter. I can do without streets of gold. Give me asphalt or concrete without potholes, speed bumps or same level intersections and it would be heavenly. At least one Christian songwriter thought along these lines. I remember a line from one of the old hymns, "Just give me a cabin in the corner of gloryland."

When I imagine the performance of the heavenly choir, I hear the Mormon Tabernacle Choir singing 'How Great Thou Art'. Of course, if it is composed of the best of all those folks I hope made it, we could hear a blend of voices from the Holder's Chapel congregation, members of the Blackwood Brothers Gospel Quartet who have gone on and Hank Williams, Bing Crosby and Arturo Caruso backed by the New York Philharmonic or Marvin McCracken's band with Lockey May playing second fiddle, the base fiddle and guitars of the Campbell family and someday well after I arrive, surely the piano and sweet alto of June, my baby sister.

On the other hand, I could never figure out where hell is and how big it would have to be to hold all those millions condemned to hell by some of our more severe preachers. I only hoped at the time there was enough room for the Germans and Japanese that were killing our boys.

Could it be that at the time of death we acquire total recall and infinite conscience. If we have been forgiven of all the bad things we have done, afterlife will be tranquil, enriching…heavenly. If we were really bad and unforgiven, our ultimate conscience would feel and total recall would remember all those bad things and become embarrassed. The heat from embarrassment would rise exponentially with the degree of badness, thus producing the 'fire' of hell. If our afterlife is spiritual rather than physical, all the good spirits could fit on the head of a pin and all the bad spirits would be impaled on the point. Or we could have all of the universe to inhabit.

One side bar to church going that I will tell now has little to do with church except the journey there. Every family has its unique sayings that have meaning only to its members. We were heading for church in our Sunday best. All of us were seated in the Model A and Daddy had impatiently waited on Mamma to get in the car and close the door so he could get us on our way. As he was backing out to leave, Mamma threw up her hands and exclaimed, "Wait! I forgot to grease the old hen." This phrase has become my family's substitute for a request to go back for something forgotten. She greased the old hen to kill the lice.

Jack continued to look for positives in the war. In his letter of August 20, 1942, he wrote, "How did you all like the raid the allies made on the French

Coast? It was a good one wasn't it? Maybe they will start on a big scale pretty soon. That will be a very rough affair when they do start a 2nd front."

He was no doubt referring to a predominately Canadian commando force of five thousand men that attacked Dieppe on the French coast. When half the force is either killed, wounded or captured, there is some question about the strike being a 'good one.' Of course, the news media with the assistance of the defense department were making an effort to boost morale by painting it as a great victory. There is no doubt that the failed attempt to invade France led to the decision to delay the planned invasion of France. (WWII, American Heritage)

Loyd was promoted to corporal and was settling in to become a real soldier. In Jack's letter of September 2, he reveals a slice of life at Ft. Ringgold and at home, indicating the comfortable state that our family had been drawn into and supported my attitude of well-being and the immunity of the Knight boys. In some way this feeling of mine could have been enhanced by my experience during the summer revival at Holder's Chapel. His letter read,

"How are you all? We are o.k. We are having it pretty easy now. We are going to school most of the time. We ride out on the trail in the morning & have classes in everything you can think of in the afternoon.'

"Boy that was a good watermelon we had last night. We took them both down to Curtis's house & ate them. The first one was just as good. We were sticking out like buttermilk pups when we got it all finished. When we got through eating we all four lay back on the bed & Joyce said 'four Knights'. Billie R. & Mary helped us eat them.'

"A little gal just passed by the window and I ask her (just to have something to say) to tell me something to write my Mom. She said to tell you that I was going to teach the soldiers not to whistle at the girls as they passed. I wonder if she thinks I'm super man! She is a 1st Sgts. daughter about 17 & very silly at times & very smart at other times. All the boys whistle & yo ho at her. She's blonde and not bad looking.'

"Curtis & Joyce just came by. They were just going for a walk on the post. It's cool, but it isn't hot as it has been. They let me read your letter. Boy you brats must be fishing demons. Did you take any pictures of it. I sure would like to have been there. Toughy write & tell me all about it. Jew didn't you think you had a whale. I can emagine how you felt. You all had to have that tobacco juice of Dad's on it tho before the bait would work. I got a letter from Buck the other day. I wrote you about it. We have got a date for 1944. We are going to get us up a ball team. When we all get together again we are all going to get drunk on Champain. You too Mama & Aunt Madell. We'll have a jubilee.'

"Buzzard is making a good Cpl. He had a hay detail this afternoon & some of the bunch left before he got through with it. He just went to the house & hunted them up and made them go back & do some extra work, & then took them to the canteen. That's the way to do it. You treat 'em rough & let them know that you aren't taking any bull & then be nice to them. They admire you for being rough with them & like you for being good to them. That would be good to try on kids. All people are just grown up kids. They don't change a great deal. Do you think?'

"June Bug I'm going to send you some money to get you a perminant. If you have any left over you can get you a dress or some supplies or anything you want. You are pretty big aren't you? 50lbs. Bill you are still a little bigger tho. You have to be so you can take care of her when she starts to school. You and Nig will probably have to whip a lot of guys buts over her before she's 18 years old. I think we can depend on you all tho. Can't we?'

"I'll give you a typical scene here at 12:00 noon. I'm coming up from the stables with my platoon in column of 2's not even in step or counting for them. When we get to the back of the barracks I holler 'fall out' & they go in every direction, mostly to the latrine to wash the dirt out of their eyes. Buzz & his bunch have been to the barracks about 5 minutes & are all sitting on the back porch & he is reading something out of the soldiers hand book about close order drill. Curtis hops out the back kitchen door & starts beating on an old wheel & yells, 'Come & get it you dam wolves' & they come."

Herchel and R.C. had caught a large catfish without Daddy's help. I recall that it weighed in excess of thirty pounds. The letter was written four days before Loyd's twentieth birthday. Curtis bounced back and forth between kitchen duty and line duty.

Jack wrote on September 10 to Daddy and the boys at home.

"Mom & them got here o.k. & seem to be having a pretty good time. They have been down at Curtis' house all time 'till today. She is out here at the post at Mildred's house. She got to see the river when it was up. It really is on a big rise. It must be raining a lot up the river because it sure isn't here. We have had one little shower. It helped settle the dust tho.'

"We are all going to the show tonight. Curtis got Mom to go to the show last night. She ought to go to more shows. They are really good for getting your mind off other things. I think she will be better satisfied about us now since she sees what we are doing."

In late September of 1942 elements of the Second Squadron were sent to Ft. Brown at Brownsville, Texas to help guard the Municipal Airport.

Jack wrote September 21.

"We have seen gillions of air planes since we have been here. They come & go every five minutes, the army & navy planes & civil air patrole & Pan-American & Braniff and Eastern. We don't guard or check any except army & navy and foreign planes but there are pleanty of them. Planes come in all the time from Guatamala & Panama & Brazil. I'm on Sgt. of the guard every other night. Most of the boys are on all the time but they are on 2 and off 4. The Sgt. seldom ever gets to sleep during his tour of duty."

The U. S. Marine Corps was in the middle of the battle for Guadalcanal and U. S. forces were preparing for transport to North Africa. War and the preparation for war were in full force. The 124th Cavalry Regiment was still stuck on the Texas/Mexico border pulling guard duty. This situation rankled Jack, as his letter of October 4 revealed. Also, he never passed an opportunity to help protect Mamma and offer advice.

"We got your card and letter today. Heck Mom I wouldn't go off up there to pick cotton. You might make a little money but what have you got when you get it. If you need any clothes or anything, I'll send you the money if you want it. You are going to hurt yourself working too hard. Besides those brats need to be in school.'

… "I think Loyd & I will go to officers candidate school. I don't know tho. We hate to leave Curtis. He won't leave Joyce. Don't tell anybody because we might not go. It would be pretty hard & I don't know if I could pass or not. We have to take a pretty stiff exam here before we get to go.'

"We will go back to the airport Saturday I guess. This is getting in my hair here. I just can't sit here while all the others are going over. I want to get away as soon as possible. If we all sat down on the border this dang war would last a million years. I feel just like a ball player sitting on the bench during a big game. I want to play.'…

"Dad how do you like how the World Series are coming out? I was a little surprised in the Cards. They really poured it on the Yanks today (Sun.) with plain power hitting. I guess they will win the series. The Yanks would have to win three straight & that ain't easy against a bunch like that." [The Cardinals won the next game and the series four games to one.]

Daddy had taken us out of school for a few days to go to a farm about thirty miles away to help him pick cotton. We camped out in an old abandoned farmhouse for about four days. For me it was fun sleeping on a pallet and eating food prepared over a campfire. I helped pick as much cotton as I could, which wasn't much. When I got back to school I expected everyone to welcome their long-lost friend, but I never made a wrinkle. After all, I had missed only two days of school. My celebrity status was only in my mind. Part of the problem was the arrival of a new boy, Joe Overstreet. His father was a welder and had moved back to the community because of work opportunities. His father subscribed to a science magazine and Joe told me that he read that someone had invented a radio with pictures. I thought he was lying. He moved away again, but came back after the war and started to school with me in high school. We became best friends.

Jack seldom had to sit on the bench for any of his teams and he found 'sitting out' the war to be irritating. His friends from home and others from Troop F who had been transferred to other units had already seen combat. He wanted to get in there and take a few swings.

Loyd had just turned twenty and was soon to be promoted to sergeant; a remarkable turn-around from a brawling, hard drinking private only six months earlier. He was beginning to mature as a soldier, but his private life revealed his lack of success in other aspects of his life. Jack wrote about his concerns on October 10, 1942.

"...I think Buzz is about to get married. It may be a lie but that's what I hear. I guess I'll just have to get me a greaser to keep in the running. It's Joyce's sister if what I hear is right. Don't think anything about it because I guess he knows what he is doing."

Events were moving along for Jack and Loyd. Jack wrote November 3.

"I guess I feel pretty good today. Maybe it's because Buzz and I are going to get to go to Tank Destroyer School pretty soon maybe. Six of us from F Troop & some from other troops went before an examining board. I made the best grade 97 & Buzz next with 96. We were the only ones that were classed Superior. Emagine that! They ask us questions on current events & common sense questions. Don't tell any of those guys around there that we are going because we may not for a long time & then we may not pass it. If we go it will be at Camp Hood between Temple & Gatesville. 135 miles from home.'

..."We are still working pretty hard. I don't know if Loyd wrote or not but you can address his letters to Sgt. Winfred now. He was made Sgt. this weekend. Curtis is making it o.k. out of the kitchen. I think he likes it better out of there."

Jack had a major change in his life coming. He wrote about it on November 20.

"How is everybody? Well I'm o.k., but feel kinda hollow. I had my tonsils taken out this morning. I feel kinda funny, like I had been dehorned. I'm getting along fine tho. I have already stopped bleeding & my throat isn't near as sore as it was this morn.'

"Murphy just came by the horsepistol to see me. They are going to have a big inspection tomorrow. I knew I got in here for some purpose.'

"I guess you want to know why I took pain all at once to have my tonsils removed. Well you see I'm going to Ft. Riley to report Dec. 18 to Cavalry Officers Candidate School, & I knew if I went to that cold country, with these old rotten tonsils I would be sick half of the time. I'd be sunk if I were sick any. I'd never pass it. I'm going to get 5 days delay in route & I'm going to hit the old man up for a little furlough to go with it. If I get it I'll be home about the 5th. If I don't I'll be there about the 12th. I didn't want to go to the Cav. school but these rats try to send guys with fair grades to the Cav. so they can keep them in this branch. They may send Buzz pretty soon. There are a lot of guys here that have had applications for Cav. for a long time.'

"I'm going to write Rowe tonight. I don't know if he will write or not. I'm going to try to get in the 10th Cav. in Cal. when I get out of school, & maybe I will get to see him. The 10th Cav. is a Negro outfit with white officers. Weyland Holt is out there & he really likes it."

Curtis wrote the next day, bragging about Loyd.

"Buzz has really been studying and has learned a lot about the army the last 2 or 3 months. We have classes in the dayroom on map reading, weapons, horsemanship, foot drill, compass reading, etc. so the other day Buzz overheard what one of the questions would be in class, so he looked in the book and got the answer, so when the troop commander, Newton asked the question, Buzz quoted the answer from the book. It really made a hit with Fig. (Capt. Newton). It wasn't long before he was a Sgt."

Loyd and Lenora got married, but hardly a month later Jack wrote in his Nov. 14 letter, "...Lee is here now but I don't think her & Buzz are getting along so good." They would divorce soon, during the time Loyd was attending Officers' Candidate School. But this love story was not over.

In a passage from Joyce's booklet we find the following account in Curtis' voice.

"Joyce and I got to Mineral Wells and Kenneth [Knight, a cousin] took us out to the house. Jack is still on leave, and we had a good time running trap lines, going fishing, hunting. We played ball and the kids got a kick out of the way Joyce could hit the ball and run. Loyd came up while we were here. We ate like hogs. Joyce got a taste of living in the country with everything we did and the good fresh pork and good food we ate. Jack killed some mallard ducks and we ate them. Dad had some sweet potatoes under a straw and dirt hill."

Operation Torch in North Africa claimed the first life of a Garner boy. Martin Troy was killed in French Morocco late in 1942. (History of Parker County)

After Jack and Loyd qualified for O.C.S., Jack began his school in late December at Ft. Riley, Kansas. Loyd would be sent to Ft. Hood, Texas for training in armored warfare. Reassuring Mamma and Daddy at another difficult juncture in his life, Jack wrote on December 19 from Ft. Riley.

"We are getting started. I may not have time to write much but you can be sure that I'm o.k. This army can't get tough enough for me. We drew our equipment today & had a lecture by our troop commander & will start study hall tonight. We have it every night 2 hrs. from 7 to 9, all except Sat. night. We have it on Sun. too." He also mentioned how cold it was in Kansas and followed that complaint with assurances that his life was just fine with plenty of good clothes, good leaders and warm barracks.

On December 27, about a week after arriving at Ft. Riley, Jack wrote,

"Well it's snowing again. It's about four or five inches deep on the level & nearly knee deep in low places. It's not too cold tho. I guess it will get pretty cold tonight. It won't hurt us tho. You should see us when we start out. We look like a bunch of Eskimos. We wear heavy over shoes & macinaws & overcoats if we think we need them.'

"I had a real good dinner Xmas. I went to Manhattan and stayed all night in the service club. (I think I told you this). We slept in cots about 50 boys. We all got invitations from people in town & some out in the country to come eat Xmas dinner. I went to a man's house that taught in the Kansas State College there in Manhattan. He & his wife had one kid about 6 & two nephews whose dad was in the army & I didn't find out where their Mom was. They were nice people & had a lot to eat, interesting too. They were worrying about the War."

Jack, always the critic, wrote in his January 1, 1943 letter,

"Curtis said they had about 75 rookies. Boy I'll bet that is a rough place down there now. I'm glad I got out of there. Yeah I like it a lot better here now. I'm kinda getting acquainted I guess. We are having weapons now. We went out & shot the M1 yesterday and will fire the light Mg. next week. Buzz didn't write. I don't know when he will come to school. I wish he would hurry and come on. I wish Curtis would come too. He would make a good officer. I'm kinda surprised at some guys they have sent here. We have a few here that wouldn't make first class private in F Troop in a hundred years. I don't see what their commanding officers meant by sending them."

Curtis had been moved from the mess hall to the line again and was taking on various jobs as platoon sergeant and stable sergeant. Apparently his most significant job was in training recruits. The army at least had the good sense to take advantage of the well-trained men of the 124th Cavalry for purposes other than guard duty. Since the war had heated up and the Allied Forces were now on the offensive, well- trained troops were needed.

During this cold winter of '42-'43, the Knight boys at home resumed another family activity that was both sport and in the past a part of the family livelihood. Jack mentioned it is his letter of January 6, "You kids must be a bunch of little Dan Boones. I never could catch a dang fox. Tried a gillion times."

Daddy had about twenty steel traps that he used each winter for trapping fur animals, which in North Texas were skunks and possums for the most part with an occasional raccoon, ringtail cat and rarely a gray fox. The lower quality of fur from animals in the South produced a low price for the furs. Competition from neighbors reduced the quantity of animals. So the venture wasn't lucrative, but provided a little more for the family. The sport of outwitting the little creatures probably had more to do with the winter activity than anything else. We couldn't play baseball and go swimming in the winter months, although later in my adolescence I overcame this cold barrier. I was duck hunting with R.C. and stripped down to go in after some ducks we killed. It was late January. It was also my last time for that foolish stunt.

Daddy had a couple of large double-spring steel traps that were set in the most likely places to catch fox or raccoons. Expecting to catch a fox, the traps would be set along a fencerow. Because the gray fox was a larger animal, it could reach a longer distance to snag the bait without advancing its front paws to the trap. To offset this ability, our strategy was to build a small triangular structure with large twigs enclosed on two sides and the top, leaving only one side open for the fox's snout. The bait was either a chunk of cottontail rabbit, jackrabbit or a songbird we had bagged with our .22 rifle. This would be wired at the back of the triangle, hoping the fox would be hungry enough to ignore the containment and dumbly step into the jaws of the trap while tearing away at the bait. In all the years we practiced this sport, I recall catching two fox. Once was this year of 1943 when Herchel and R.C. got lucky and again in the winter of 1950-51 when I got lucky. In our environmentally insensitive minds this was the ultimate in Dan'l Booneness. I have to admit that it was not a pleasure to use my .22 rifle to shoot the frightened, cowering and injured animals before taking them from the trap. However reluctant, my desire to make a few bucks and to establish my place in the manhood society motivated me to move on and gather my meager catch and process the furs.

The set for a raccoon was much different. The flat trigger mechanism would be wrapped in tinfoil and placed at the edge of water in a pond or stream. Stink bait, made from rotting mackerel, would be sprinkled about. The 'coon was expected to be attracted by the stink, see the tin foil 'fish' and grab it. I suppose that if this tactic ever succeeded, that all the really dumb raccoons had been caught. It didn't work for me at all; not even a sprung trap.

I will spring forward a few years to my adolescence. My first effort to trap on my own was also my last. During the Christmas/New Years vacation of my senior year in high school, I decided that I needed to make some money. The wartime economy had improved the family's financial condition and Daddy had given up on trapping altogether.

With little regard for history and with visions of fox, raccoon and mink, I embarked on a two-week quest for riches. The scarce and elusive mink was always the end-of-the- rainbow. We never caught one and no one I knew ever caught one, but the folklore of the possibility of a $100 catch was passed on year-to-year. A more realistic fantasy was the catching of a $15 raccoon, a $10 fox or a $5 narrow striped skunk. I gathered up all the rusty traps, checked their mechanics and sanded the edges of the jaws so they would snap correctly and took my .22 and the jar of rotting fish along with the traps and began my circuitous route through our farm and adjoining farms. We never asked permission, it was just a given that it would be o.k.

Having followed Daddy and my brothers on these expeditions I had a sense of where to place the traps. On the initial trip, I spent hours searching for the perfect trap sights, arranging the environment to the most enticing and deceptive condition and moments to admire my handiwork. I ran the trap line early each morning. I had moderate luck in numbers, but phenomenal luck in snagging a fox. I could barely believe that I had joined that elite group of trappers when I saw the frantic animal. I was only seventeen in my first independent venture. The sky was the limit. Who could deny that I was a premier outdoorsman. After all, I had accomplished what few others had, including my idolized Daddy and older brothers.

I spent hours skinning, stretching and scraping those hides. A couple of 'coon hides and assorted possum and skunk pelts had been added through some night hunting with our dogs and carbide spot light. They were easier to catch with dogs than with traps. We had to hunt in territory away from the trap line to avoid the possibility of trapping one of our dogs.

Soon after school resumed, on a Saturday, Daddy and I took these prized pelts to the fur dealer in Mineral Wells. It didn't take long for this squinty-eyed, greedy, unsympathetic, soulless, dirty, stinking flesh peddler to shatter my expectations of riches. Instead of the fifty dollars I had quiet conservatively estimated the thirty hides to be worth or the fantasy of $100, he informed me that the combination of the poor quality and suppressed market, that he could offer only $15. Having no recourse, I accepted the money, my fate and the lesson. The twenty-five cents per hour for very unpleasant work was another message to me that higher education should be my 'trap line'.

Our experience with furry animals wasn't limited to trapping. On one of our outings into the woods for recreation with the dogs we were successful in capturing a fur animal alive. Ted and Bruno had chased a squirrel up a hollow tree. The entrance to the hollow was about fifteen feet above the ground. To get at the squirrel we had to cut the tree down. We wanted to capture the squirrel alive and place it in a wire cage we had at the house. This meant another trip to the house to get an ax and the crosscut saw. After clearing away the briars and brush at the base of the tree, Herchel and R. C. cut down the tree with the saw. Junior's job was to stuff his wadded-up shirt in the hole as soon as the tree hit the ground. He was successful and the squirrel didn't escape.

The boys worked together to cut short sections of the trunk half way through and with the ax split a small piece away from the tree trunk. We gradually worked down the tree, each time stuffing the shirt in the mouth of the remaining cavity. At last we saw the squirrels feet, but unfortunately, they had become detached from its body. We had to solve the mystery. After another section was cut, we carefully raised the cut section to discover a whole animal. However, it was not a cannibalistic squirrel, but a weasel-like ringtail cat that had at once captured his lunch and reaped revenge upon this home invader.

What luck! These were relatively rare little animals whose pelt could bring a decent price on the fur market. We were late into the spring and long past the fur season, so the decision was made to place it into the cage. We were excited about the capture and looked forward to keeping the ravenous little beast without any regard to how we would keep it fed and unclear of its future use.

Early the next morning, the first thing Daddy did was to check on the ringtail's condition. It was gone, having pushed through between the net wire and the frame of this loosely constructed prison that had been designed to transport chickens. The whole family was disappointed, but rationalized that it was for the best and that it removed the responsibility of caring for our fledgling menagerie.

During January, February and March of 1943, Jack's letters concentrated on describing his training, expressing concern for Loyd and encouraging the kids at home. His training was much the same as he had already experienced in weapons and horsemanship, only more detailed and intense; the distinction being training in leadership, teaching and scheduling. Included in all of this was the underlying implications of becoming an officer/gentleman.

He was very concerned about the young Loyd who had impulsively married Joyce's sister, Lee, only a short time before he went to O.C.S. Apparently, the marriage was on shaky ground from the beginning, but his absence was the last straw. The marriage failed and this had a negative effect on Loyd's pursuit of a commission. He overcame his depression and reluctance to make it through, received his second lieutenant bars and was assigned to the 519th Tank Destroyer Battalion at Camp Hood. I was very impressed with his unit patch showing a black panther crushing a tank in its jaws.

Another indication of Jack's frugality came to light in his January 31 letter.

"I've got a lot of clothes now. Boy I'm going to look like a fool with that queer uniform in places where they don't have cavalry. I don't care tho. If they don't like it I'll just tell 'em how much it cost. I think I'm going to have to spend all the $250 they are giving me for clothes. Uncle Bill Smith thought at first he could make about 50 or 100 bucks off the deal, but I am about to decide that it's impossible."

In making light of his own parsimony, he invoked the image of Bill Smith, at one time a neighbor of the Knight family. He was widely known for his ability to make the most of his resources. Curtis told me a story on this subject. Mr. Smith had hired Jack and Curtis in their youth to haul in his baled hay. Each was to earn $1.00 for the day's work. At the end of the agreed upon workday, rain clouds were gathering and he asked the boys if they would stay to haul in the last bales before the rain came. They agreed to do this and worked several hours until dark to complete the job. As they finished, Mr. Smith gave them $1.00 each. Curtis said that Jack didn't like it and let Mr. Smith know that they were due more money. At that moment Mrs. Smith came out near them and Mr. Smith yelled to her, "Bring these boys another dime!" With that, Jack gave up and took what he could get.

We bought coal oil from Mr. Smith for our lamps and to help start fires in the winter. Curtis had the job of going to his house to make the purchase. On one occasion, Mr. Smith let the can run over, spilling a small amount on the ground. The next time Curtis went for oil, Mr. Smith told him that he didn't fill the can full to make up for what he spilled the last time. Times were hard during the depression, but not that hard.

These events occurred in the thirties during the time we lived on the Crosthwaite Place. The house was located on the highest elevation in the neighborhood and because of the robust activities around the Knights' house, it was referred to as 'Noisy Knob'.

By February 9, 1943 all of the Japanese on Guadalcanal had been eliminated by the Army's 24th Division.

Curtis wrote on February 26,

"We left Laredo going back to Ringgold and got orders for Sgt. Baker, Lt. Pierson and I to go to Ft. Bliss to school for about 3 or 4 days to learn a certain phase of commando raid training. Three men from each troop were sent. Then we will have to go back and teach it to the rest of the Regiment: so that means we will be going down to Ft. Brown for a week or two when we get back I guess."

On March 2, near the end of his school, Jack wrote,

"Well eight more whole days to go. I hope I make it. About 10 or 12 of our class flunked out so far & they aren't thru yet. I'm holding my breath. I've got good grades tho & no failures in any of my subjects. I guess that will help & I can ride pretty good too. They might keep me & let me finish. Some of the ones that failed thought they were passing up till today." The army has a way to create doubt in the very best.

This letter invoked a thought of my first experience of real injustice aside from the hazing from my brothers. As most youngsters, I had a strong sense of justice. Daddy let me down twice during these years. I can understand in retrospect his decisions, but at the time one made me furious and the other deeply disappointed. After surviving a decade of fiscal deprivation, the healing economy brought some money into the family. Daddy had a steady, money-producing job at Camp Wolters and The Boys had steady incomes, however meager, as soldiers.

During one of the times when all three boys were home on leave, Jack had the idea that the five of us at home needed motivation to do better in school. Their solution was to bribe us for making good grades. We were to receive twenty-five cents for an A, and ten cents for a B. Lower grades weren't considered worthy of reward. The Boys were assigned to Herchel, R. C. and Roy. Daddy was to pay me and Mamma was to provide June with her academic largesse. When we got our report cards, all the other kids had earned less than a dollar. I had earned $1.25 for all A's. Word was sent to JackCurtis'nLoyd and soon the money came. Mamma paid June from her egg money. Daddy recognized the folly of giving an eight-year old $1.25 and simply refused to pay me. I think he realized I had been making A's all along and didn't need extra motivation.

Actually my success was unfair to the others. The year before, Texas had gone from an eleven-year program to a twelve-year program and had allowed students to skip a grade. I had started first grade at age five, this creating a situation that had me entering fourth grade at age seven. I also suffered the agony of leaving the secure 'dog house', a separate building where the first and second grades were housed. I was depressed, lost and overwhelmed by my new surroundings. Soon after school started, I feigned illness and the principal took me home.

Mamma's insight saved me. I give her credit for the impetus to my academic career that would eventually culminate in earning a Ph.D. in educational administration. The other kids had remained in the grades they had been placed and except for R.C., proved in the end to succeed. But that first report effectively ended that attempt at rewarding good deeds. My good grades were more attributable to the curriculum, which was to a large extent on my maturity level, rather than my academic brilliance.

Rationing of certain scarce products and commodities was a fact of life. I recall gasoline rationing with the windshield stickers signifying the level of one's status in the war effort. I also carry an indelible memory of sugar rationing that caused Daddy to commit another lapse in fair play.

With our newfound affluence we could occasionally afford to buy corn flakes for breakfast. This was much more expensive and better tasting than oatmeal. With seven hungry mouths around the breakfast table clamoring for their bowl of corn flakes and several heaping spoons of sugar, supply of both was constantly being strained. This led to the squabble over equitable distribution. My parents solution to this was to divide the sugar ration into 'cooking sugar' and into equal amounts of 'personal' sugar, kept in individual coffee cans on which our names were inscribed.

My genetic disposition to hold back part of my possessions for the future led to another crisis. Daddy loved large amounts of sugar in his coffee, iced tea and cereal. Of course, his supply was depleted long before mine. Naturally, he thought it only fair to take from the fullest container. I will never forget the fury that engulfed my being on discovering that my carefully hoarded supply of sugar had been looted, not by my aggressive, insensitive brothers who could care less about how I felt, but by Daddy, the man I thought was a special person. After all, he regularly left something for me in his lunch pail. I don't know now and I doubt that I had a clear notion then about what I was going to do with that hoarded sugar.

My protestations took the form of quiet tears and looks of betrayal. Well, this contrived effort at fairness was abandoned to the more honest practice of the survival of the fastest. I usually lost and learned to drink tea without sugar and eat cereal with a little less.

CHAPTER 10

A MAGICAL DAY WITH JACK

DURING THE SPRING and summer, 1943, allied forces continued their successes against the Japanese. Command in this area of operation had been divided between General Douglas McArthur and Admiral Chester Nimitz, not for military purposes but to assuage the egos of these brilliant and prideful militarists and their respective branches of service. McArthur was assigned the task of clearing the Japanese from the New Guinea coast up to the Philippines. Nimitz , Commander of the Pacific Fleet and U.S. military forces in the Central Pacific theater was made commander of the Southwest Pacific Theater. Nimitz ground forces consisted of U.S. Army and Marines. McArthur had U.S. Marines and Army plus Australian Army divisions along with air support from the Fifth Air Force and the navy under Admiral Halsey when needed.

On March 3, the Fifth Air Force under General George King attacked a Japanese convoy enroute to New Guinea with some seven thousand troops to reinforce their operation in New Guinea. The attack was highly successful and so devastating to Japan that it would be the last attempt at a major troop movement by sea.

Nimitz and McArthur devised the very effective strategy of island hopping. The Allied Forces would skip the more heavily fortified Japanese-held islands and invade those less formidable. With the recently won naval and air supremacy, the stronger islands were isolated and left to starve, assuming an inconsequential role. Even with this strategy, the fight was savage and for the Japanese, suicidal. We can only speculate about losses to both sides if we had, as the Japanese expected, taken every island occupied by the their forces.

The action in North Africa finally swung in favor of the Allies. After taking a beating by Rommel's forces, the U.S. Army pulled back and gave their soldiers more intensive training with new leadership. Our air force sunk much of the Nazi shipping to deprive their army of much needed supplies. As a result, Axis troops were cleared from North Africa by May 1943. (WWII, American Heritage)

Curtis, now fully involved in troop training, gave more details of his own training in commando tactics, "I'm still dodging bullets and crawling through barbed wire about 8 inches high and under machine gun fire about 12 to 15 inches over our heads. It's safe if you have sense enough to stay low. No one has been shot yet, and a lot of guys have gone through it. We've been throwing hand grenades, judo and quick firing of all our weapons, including machine guns-without aiming. We are using all live ammo." This was another bit of information that heightened Mamma's anxiety.

Jack graduated from Officer Candidate School and came home on leave. The uniform of a U.S. Cavalry officer in 1943 was enough in itself to strike awe in my nine-year-old psyche. From the spit-shined, leather riding boots with the English spurs, to the pink riding pants, the olive green jacket with leather straps and brass buttons and hat with leather bill and officer's insignia, the effect was total majesty.

Jack had been commissioned as a second lieutenant. I knew all the military ranks of all the services in their proper order. That he was only a second lieutenant did not diminish his magnificence in my eyes. When you're barely four feet tall, a man just under six feet and weighing one-hundred ninety pounds with broad shoulders and incredible strength, with a precision cut flat-top hair style was more than a brother.

He was home for a few days. I had to be with him, all by myself, for at least one day. Each day the five remaining kids walked the mile from our weather-beaten, unpainted house to the bus stop to ride the remaining two miles to school at the tiny village of Garner. On this day I would miss the trip with my brothers and sister. I was not a habitual liar, although I had learned that I could: maybe a good actor, but not a liar. Today I had to find a way out of going to school.

I was awakened by the usual sounds of Mamma and Daddy getting breakfast ready. He would be leaving shortly for his civil service job at Camp Wolters just across the lake from our house. He had to drive about five miles around the lake to the military post that had become one of the largest army basic training camps in the country. On a still morning we could hear the sounds of drill commands drifting unimpeded across the lake, another part of our emersion in war and its preparations.

The house was our typical abode during these years, being constructed from unpainted pine boards. The interior of the three room shack was papered with a product from the hardware stores in Mineral Wells that my parents hung over other paper hung by yet others who lived there before us. All of this was covered by a roof of sheet metal that amplified the sound of every drop of rain or hail stone that fell. The house faced east and the boys slept in the north room. Daddy and Mamma slept in the south room where the wood-burning heater was located. Their room served as the living room also. June's bed was in there too. The small kitchen and dining room was located off the south room to the west. The first discernable words of the day came from Mamma.

"Roy, stop that!"

The exclamation could have been in response to a flirtatious poke or pat, but was probably a reprimand for one of Daddy's many annoying little habits that he had cultivated to aggravate Mamma. There is no delicate way of saying that Daddy was no doubt the Parker County master of the early morning whistling fart. Another day had begun in the Knight household, but not an ordinary one.

"You boys get up and keep quiet. Don't wake up Jack," Daddy whispered as he gently shook each of us to make sure we were awake. I had been lying there awake trying to formulate a scheme that would allow me to stay home from school that day. During those years, I went to bed about eight o'clock and was ready to get up early.

After getting into my overalls and brogans and joining the others at the table in the kitchen for breakfast, the idea began to take form. I was never a great breakfast person anyway. The greasy fried eggs with the brown lace around the edges from frying too fast in too hot lard were easy to forego.

"I don't feel like eatin'," I announced.

Mamma, who was very protective and solicitous to her kids, bought it.

"Are you sick, Bill?"

I knew I had made the first hurdle and decided to go all the way with it. Somehow I knew that it would work because I was a pretty good country scholar and always wanted to go to school. Besides, Mamma might have known the real reason I wanted to stay home. She was very intuitive, as good mammas are. I went back to bed and waited.

The older visiting brothers could always sleep late, which was a source of irritation to me because their visits home were so infrequent that I felt it was a waste of time for them to sleep so much. After Daddy left for work and the boys and June had time to catch the bus, I began to feel much better. I was ready for the day. There was no guarantee that Jack would be around. He had friends in Mineral Wells and around the countryside and Daddy rode to work with a neighbor, leaving the Model A Ford for him to use. Jack finally awoke and came to the kitchen table.

"Honey, do you want a cup of coffee?" Mamma offered.

"Sure. I slept like a log. It sure is good to be home. Hey Cotton Top, what are you doing home?"

"I was sick this mornin', but I'm OK now," I answered, a little contrite.

Jack grinned and touched my shoulder. Mamma, always ready to serve, asked Jack what he wanted for breakfast.

"Oh, I'm not all that hungry, but I could eat a couple of fried eggs and some ham, if ya'll have any left. Warm up some of those biscuits, too."

As she eagerly went about the business of Jack's breakfast, she began to probe,

"Well, now that you're a lieutenant, what do you think they're goin' to do with you?"

"My orders are to go to Fort Brown in Brownsville. I'll be at the 124th headquarters. It's right on the tip of Texas close to the Gulf on the Rio Grande River."

"I am so glad they aren't sending you over."

"Mamma, I know how you feel about us fighting, but you better get ready. Our time will come. Are you still worried about us if we get in a war?"

"Lordy, yes! I just don't see why we can't let those Japs have their part of the world and let them countries in Europe take care of their business."

"We can't let those Nazis and Japs have their way. Next thing you know they'd take us over. We can't let 'em do that. Besides, the Japs need to be punished for what they did at Pearl Harbor. I want my chance to help out and I'd rather go after the Japs."

"Oh, honey, don't talk like that. I cain't hardly stand it."

As he began to eat his breakfast, Jack asked, "Are you still worryin' about us having to kill somebody?"

"I've always worried about that, but not quite as much as I did. I prayed and prayed about that and couldn't get any comfort. I listen to that preacher on the radio out of Wichita Falls and I wrote him about it. He wrote me back and said, according to the Bible, there are times that it's OK to kill. It's hard for me to see how the Bible can say, 'Thou shalt not kill' in one place and in another place say, 'There is a time to kill.' Maybe them old Jews were talkin' about two different kind a' killin'. Anyway I'm not as worried about our boys' souls, but I still don't like it."

"Well, I've got to tell you. I won't mind at all killin' some of those sneaky Japs." [In his mind all Japanese were guilty in the same way that all Mexicans were still responsible for the Alamo.]

"I just hope this war is over before you boys get into it." Mamma wasn't easy to dissuade.

Jack had had enough of this conversation and as he finished his last biscuit with peach preserves, he looked at me. I was hoping really hard that he would stay home so that I could listen to him tell Mamma and me more about the army. I wanted this to be a special day. My day.

I suppose anyone with feeling reacts positively to adoration. Jack must have known what I needed. I needed to be the center of his attention for at least this small time. When you're the youngest of seven sons and with a Shirley Temple sister, your chances for the spot light are reduced immeasurably. But he knew.

"Bill, let's go duck hunting."

"O.K.!"

I couldn't believe my luck. Since I was too young to have been taught the use of Daddy's twelve-gauge Remington pump shotgun, that invitation could mean only one thing. I would follow him while he did the hunting. It was mid-morning when we left the house. Ted, the Collie, and Bruno, the white, black-eyed dog of unknown ancestry tried to follow, but they were ordered to stay. They were not retrievers and would be in the way.

We crossed the goat-wire fence of the City Place farm and walked through an abandoned field. The city of Mineral Wells owned this property and it had no tenant who farmed it. As a result the fields were grown over with mesquite, bluestem, fescues, Bermuda and Johnson grass. There had been warm days in February. The first alarms of spring had sounded and a mat of green had responded to form a tinted base for the brittle grass stems from last year's growth. We could hear the commands of the sergeant giving close order drill commands drifting across the lake from the Negro area of Camp Wolters in his deep baritone, musical cadence.

We hit the rim of the hollow about two hundred yards north of the house and followed a seldom used path between the post oaks, black jacks and red haw as it meandered among large sandstone boulders covered with lichen and algae. A few white buds of the wild plum had overcome the restraints of winter to emerge into the sunlight and courageously show their beauty. The path dropped off sharply toward what had been the creek at the vertex of the hollow only twenty years ago, a few yards from the site of Herchel's conquest of the armadillo.

It was near this trail that I witnessed one of Herchel's feats of 'toughness'. One of our more often practiced Sunday afternoon activities was to take Ted and Bruno and head for the holler to hunt any varmint that the dogs could find. The dogs had to maintain constant vigilance for snakes. The most prevalent species on our farm and in the holler was the copperhead, less venomous than the diamondback rattlesnake. The dogs were usually bitten about the head as they tried very cautiously to grab the serpent and shake it to submission. I never learned why, but Daddy's treatment for these bites was to force feed lard to the dogs. Their heads would swell to double size and they would lie around the house for a few days recuperating and pondering the wisdom of their aggression.

The animal most plentiful was the armadillo. We had no use for the prehistoric beast, but they presented a challenge. The dogs had run one into a hole. We couldn't reach it and it could dig faster than the dogs, so R. C. & Roy went to the house and brought back a shovel and grubbing hoe. Herchel and R.C. began digging through the sand and rock. The task was difficult, but they finally saw the tip of the armadillo's scaly, tapered tail. Herchel wanted the armadillo alive. He squatted in the ditch that had been excavated and grabbed the tail and began to pull. The natural response of the armadillo was to expand and contort its scaly body against the walls of the hole and dig its prodigious claws into the cracks of the rocks. Few men, if any, could have the strength to overpower this combination of resistance, but Herchel had a solution. He would keep the pressure on and wear him out to physical exhaustion. After what seemed hours, he was finally successful. He pulled the expended old warrior from the hole with a very satisfied countenance. I could hardly believe it, even with the seeing.

The victory was simply too sweet to ignore. Surely there was some value to this critter. Herchel and R.C. began a discussion about eating it. They had heard that fried armadillo tasted like fried chicken. We headed to the house with hunger in our eyes if not in our bellies, to be disappointed. None of us, not even Daddy, had ever dressed one to eat. They had to shoot it in the head to kill it, but that was the easy part. Trying to remove the tough skin of the underbelly and the armor plate on top, for neophytes, proved impossible. After a lengthy attempt to peel the hide and shell from the corpse, they admitted defeat and threw the bloody, mangled mass of flesh, bone, and armor off the cliff near the back of our house. I have never tasted armadillo and have never felt deprived.

Jack and I didn't have to worry about the copperheads and rattlesnakes as we continued our journey to the lake. It was March and cold, and after a few days of being teased into the sunlight, the snakes had again disappeared into their subterranean cells. In another month the place would be infested with them.

After the dam was built to form a reservoir for a water supply for Mineral Wells, a good fishing and hunting lake was formed and the shoreline was about halfway down the incline. We approached the south side of the small cove behind a lush growth of cattails. There was a slight northerly breeze, which was very chilly as it left the water. The direction of the breeze worked to our advantage, carrying sound and scent away. I said it was my day.

"Be real quiet and stay behind me," Jack whispered as soon as he saw that a small flock of Mallards, Red Heads and Pintails were feeding along just off shore away from the cattails and in a clearing amongst the lily pad.

He maneuvered us into position where there was a small clearing in the cattails. The ducks were drifting east and would soon be in the open for a clear shot. The first shot would be taken with them on the water. We were accustomed to hunting for the table, not pure sport. I had been taught respect through fear of firearms. Daddy forbade any of us the use of his favorite gun without reaching the proper maturity and without detailed instructions. The shotgun was the gift from the Dallas hotel executives who had a small lodge on the lake and who had hired Daddy to guide for them and build blinds for their duck-hunting trips.

As the flock moved into an advantageous alignment, Jack fired. I was almost on top of him, looking over his shoulder. With the recoil, the noise of the gun and my adrenaline flow, I was a heart-throbbing mass of excitement. The shot wasn't perfect, but his first shot left three dead on the water. With quick reflexes, Jack managed to knock down two more as the large, fat, beautiful birds made their frantic, but cumbersome bid for escape. Five ducks were on the water and we had no dog and no boat. Patience was the solution, because we never had retrievers. We let the wind do our work. It would take some time for them to drift to shore and then the lily pad would catch some of them. Lapsed time was bothersome because military lake patrols could interfere with our retrieval at any time. I don't recall the conversation but it was filled with instructions to me about what we were to do. I remember chunking rocks and pieces of fallen trees to dislodge the ducks and increase the wave action to force them in. Jack finally had to remove his boots and pants and wade into the water to retrieve the more reluctant floaters. He dried himself without a fire to avoid detection and dressed. We had collected five ducks, two with that magnificent, iridescent green found on the heads of Mallard drakes, one Redhead drake and two hens.

We made our way back to the house. Mamma joined us in picking off the feathers. She saved the down for use in pillow stuffing. Mamma had plenty of experience in roasting ducks with a blend of butter, onions, other vegetables and spices, in the oven of the wood-burning range. Roast duck for supper. That was one of Mamma's tastiest culinary specialties.

After a dinner of some of the last sugar cured ham, home fried potatoes and canned green beans, along with a pan of cornbread, Jack decided he would drive the Model A to the Garner store to visit and to get the groceries Mamma needed to complete the supper meal. I wanted to go and he let me. He changed to his dress uniform. On the way we crossed Rock Creek over a rickety old iron-framed bridge with a plank runway. It made a big racket as the car rumbled across. A few yards south of the bridge in the bend of the creek was located Blue Hole, our favorite swimming hole. It had clear, running water, unlike the muddy stock tanks on the farm.

We were almost halfway to the store before I realized the potential dangers of the trip. The store was in sight of the school. What if my teacher saw me, or even worse, the principal? The closer we got to the store, the more anxiety I felt. Maybe it wasn't my day after all. We approached the town from the west, the school in view from the south and on my side of the car. I tried to disappear. As we stopped I was glued to Jack's side. I slid across and got out on his side. With luck I wouldn't be seen going in if I kept Jack between the school and me.

The store building was constructed from native sand stone and had a drive under the porch in front with gasoline pumps under the outer edge of the porch. I discovered in the summer of 2001 that it still stands and houses a sheet metal business. The gasoline pumps were the type that had a twenty-gallon, clear glass tank at the top, into which one had to hand-pump the amount of gasoline desired. Gravity would then force the fuel through the hose into the gas tank of the vehicle.

As we made it through the door, I felt like I would be safe, but inside the store were at least a half-dozen men, including Wade Howard. Mr. Howard and Josie were the parents of Rowe, from whom I acquired my middle name. Our families were friends and Sunday visitors. He had always liked to tease me. He was joined by some other experts in the art, members of the 'spit and whittle club.'

"Billy Rowe, why aren't you in school?"

"You better watch out. Miss Trammel will see you."

I was speechless, the perfect target to tease. I was a gullible, tow-headed, freckled, cherubic kid with light skin that glowed blood red in embarrassment or anger.

"I think I see Mr. Kell coming this way."

I said nothing, but pushed further back between Jack's legs as he sat facing the cast-iron wood-burning heater with an open front hearth that the really good spitters could hit with their tobacco juice with precision. I tried hard to become invisible and keep as quiet as possible, because I was naïve enough to believe every word. But I was just a passing interest. Their real interest was in Jack.

All of them knew he had left as a private in 1940 with Curtis and Loyd. They also knew by his uniform that he was now a commissioned officer. They wanted to know about the Knight boys and how they fit into the war. After an hour of discussing the war and all the local boys who had gone into the various services and which ones wouldn't be coming back home, we got the groceries and headed home.

I didn't think of it at the time, because the concept of war was still undeveloped in my frame of reference, but this bothered Jack. I knew a little more than I did when they left in 1940. Then I realized only that they would be riding horses and being in 'calvary' and that I wouldn't see them too often. Now there was war and I had heard about Martin Troy being killed in North Africa. I knew about death and I was beginning to develop the unwanted notion that my brothers were subject to be placed in a position that could threaten their lives. Even though I had decided that death wouldn't find them, my faith had been rattled.

As we got closer to home and further from the school, I began to relax and believe that I had actually made the day without something dreadful happening. I had no thoughts of tomorrow. As we turned off the main gravel road onto the dirt lane that led to our house, Jack reached into his jacket pocket and handed me a stick of peppermint candy. It was my day.

Jack couldn't resist teasing all of us about whatever he thought would aggravate us the most. The next day just before supper, he started teasing me about my girl friends. I didn't have one, but that didn't matter to him. He knew the names of some of the girls my age. My face burned red and I started to cry. My hero was making me feel bad. The feelings became so intense that I thought the only way to escape the pain was to leave home. I started down the sandy lane toward the main road. My anger and hurt were in command the first hundred yards and then my concern for survival interceded. How could I get supper? Where would I sleep? I could catch a rabbit or kill a quail, but I had no gun. I could catch a fish, but I had no hook and line. I could eat with Granny Knight or Uncle Willie, but they lived in Mineral Wells. I could never reach them before sundown. I didn't want to be out on the roads or in the woods in the dark. I was just a little kid and really scared.

At about two hundred yards, still in view and within hearing of the house with my head down and my progress slowing to an ant's pace, I heard the most welcome and relieving sound of the day.

Jack yelled, "Hey, Bill! Supper's ready!"

I turned on my heel and picked up speed to the house. Nothing more was said. I guess Jack had had a chance to orient (threaten) my other brothers. His teasing ceased. We had supper amidst the usual chatter and story telling. Jack stayed a few days more and returned to Ft. Riley to await assignment while taking more training.

He wrote March 23, 1943,

"Boy we have good eats now. We eat in the Officers Mess & they put on the big pot. I'll bet I get fatter & fatter." Jack enjoyed, but would never fully embrace his newfound world of perks and privilege.

Curtis wrote from Fort Ringgold on April 5.

"We have really been going and blowin the last 2 weeks. It's been after dark when I get home every night since about the 24th of March. We've been on the rifle range all day. It wouldn't be so bad if I could shoot everyday, but I just get to shoot about 2 days and have to coach the other guys the rest of the time. I needed 4 points making expert with the M-1 rifle. When we had the old bolt action 1903 rifle, we could all shoot a lot better, but we'll get used to this one tho."

Mamma's worrying about the boys was always with us. Apparently she had complained to Jack about not hearing from Curtis. He answered in his letter of April 5.

... "You should know Curtis is o.k. If he isn't, don't worry they will send you word sooner than you could find out yourself. I think I'll break you from that stuff. We've got you spoiled writing so much. I'm glad we can tho. We may be where you can't find out about us when you want to."

Of course, Mamma never stopped worrying and Jack never stopped writing. Eventually they found themselves out-of-touch in their only overseas assignment.

Jack was reassigned to the 124th Cavalry, but at its headquarters at Ft. Brown in Brownsville, Texas. His letter of April 19, 1943 included this paragraph.

"Here I am out in the brush. We just got into Brown in time to move out with the outfit. We are 21 miles northeast of Hebronville. I think the Ringgold boys are coming out tonight or tomorrow. I'll probably meet up with Curtis out here. I hope so."

"I'm in C Troop & have the M.G. platoon so you see I'm happy again. This is my old job."

Later, on April 23 he wrote,

"We are all fine. I mean my boys & I. I sure like my troop & the Tr. Commander. Most of the other officers are young and ignorant like me so I'm not alone.'

"We had a very secret job to do all week. We left out the morning after I got here to Brown & came up to Hebronville & then on up here (Laredo). We guarded the railroad coming out of Mexico here at Laredo. Curtis was on down the track about 13 miles but I couldn't see him. We had the town of Laredo & a little piece east & E Troop started in & went for 13 miles & then F Troop was next. I sure wanted to see him but never had the chance."

This operation was called to guard the railroad during a presidential trip when President Roosevelt went to Mexico for a conference. Curtis wrote about it on April 25.

"Got back from guarding the president. I was standing by the railroad by F Troop HQ when his train came by. It wasn't a fancy train so nobody knew it was his train. They told us when it would come by. Buzz was here today about noon. We are going down to Brown to see Jack." He wrote again on April 27.

"Sure was good to see Buzz and Jack again. Jack told Joyce that he didn't want her to let me ever be an officer because she would have to be dancing with some Colonel to get me promoted to Captain. I guess he has seen some of that at Ft. Brown. Col. Shea has been trying to get me to go to OCS, but I'm not going, it would be another separation from Joyce."

Loyd made it through O.C.S. and graduated on April 23, 1943, a month after Jack and one year after his incarceration. After their return from Laredo, Jack wrote about seeing Loyd.

"I guess Lt. Loyd W. is home by now isn't he? We really had a nice little reunion. That Pot Guts & Joyce are just as bad as ever. Buzz looks like a General doesn't he?"

Again on April 28 he wrote teasingly,

"I guess you are enjoying Loyd's stay. I wish you could have stayed longer here Buzz but I know Mom wanted to show you to all the Ballew Springs socialites. She wants to make them jealous. I guess you have been up to Howard's. When have they heard from M/Sgt. Rat R. Howard. I guess he is too busy to write to them much."

Jack had some anxiety tempered with pride in being an officer out of Troop F. He mentioned this in his May 4 letter.

"I'm pretty glad right now because we are going up to Ringgold Wed & stay till Sat. I'll get to see all of the old gang again. They are going to laugh at me I know. I'll bring them to attention if they do. I'm going to get Scrony Aaron on the line."

In his May 9 letter Jack mentioned the family's impending move from the Howard Place and homesickness for F Troop.

"I hope you all can get a place where we can send you telegrams and phone you once in awhile. I would call every once in a while if I could. Do you think you will get that place close to Youngbloods? If you can find a place for sale for around a thousand dollars take it, & we'll have us a place of our own. I got $164.00 this month. I'll save about $100 of it I guess.'

"I saw Curtis play ball the other day. They played B Troop & got beat 3 to 2. That Pot Guts is the best 3rd baseman in this valley. I'll bet he made 12 or 15 good plays without an error. They haven't got a pitcher that can do any good tho. I wish Buzz & Shorty Holt & I were still up there.'

Camp Wolters was increasing their territory to the east so that it could expand its training capability. Its new territory included the Howard Place. Jack was encouraging Daddy to buy our own farm, but we leased a farm on U.S. Highway 180 between Weatherford and Mineral Wells close to Holder's Chapel and in the Millsap School District. The farming land was poor. Herchel would be gone and Daddy decided to forego farming while we lived there for two years.

Jack liked to brag about how tough things could get and his ability to take it. This was no doubt to let Mamma know the army couldn't hurt him. His letter of May 13, 1943,

"Gosh I hope Kenneth [our cousin] don't go over so soon. Heck he don't know what it's all about yet. I guess he is learning fast tho. If he & Loyd go across I'm going to get into something where I can go across. I've been loafing long enough.'

"We have been having some good stuff today. It was the Army Ground Force Physical Test & its really a test. We had to do 33 pushups, run 300 yds. in 45 sec. & run 75 yds in 20 sec with a man our weight riding our back & then a lot of other stuff worse & to top it off a four mile hike in 50 minutes. It's really good stuff.'

"I got a letter from Curtis today. They are o.k. He has a new Troop C. O. & he is tickeled pink. They all hated Fig Newton. They will know later tho that he was a man & no cream puff soft soaper.'

He wrote on May 16,

"I heard from Curtis this week. I got a letter & then had a lot of guys to say they saw him. One Lt. went up there on a test team & heard Curtis drilling some men and knew he was a Knight & had never seen him before. He said he sounded just like me. We do sound a lot alike I guess."

Jack continued to be frustrated with his lack of contribution to fighting the war. He wasn't satisfied with the essential duty that he now fulfilled. He didn't want to train men to fight. He wanted to lead them in the fighting. He wrote May 25, 1943,

"How do you like my stationary? Lt. Rierson from F Troop gave it to me. He & several of the Ringgold men are here taking riding lessons. Boy I'll teach those Parker County watermelon cowboys how to ride when I get back home. They will teach me to fight I guess tho. I'm flat ashamed of the little effort I'm putting forth to win the war. I'm just doing things that you have to pay to do in civilian life. If Buzz goes across I'm going to the Air Corps."

This is not the only time Jack spoke of the Air Corps. He had a sentimental attachment to the horse cavalry, but was he fascinated with the thought of flying? Was he merely anxious to get into the fighting part of the war? I don't know. I only know that his younger brother, Roy, found his way into the U.S. Air Force and would emulate Jack. Roy's flying legacy continued with his youngest son Bryan more than fifty years after this letter from Jack as he demonstrated that little changes with the passing of generations.

Again on June 2, Jack writes,

"We are still riding every day. I like it o.k. but that's all I do like about this place. I can't decide wheather to stay here or put in for the Air Corps. I've got to get into this thing some way and I'm mighty afraid I won't ever do it in this outfit. I hate to get out of this tho because it is the last horse Cav. Regt. in the U.S. army with white people in it."

He continued his lamentation in his June 7 letter,

"I got a good letter from Porter McQuery today. He is in North Africa & they don't care for him writing all about it. He said they were having a good deal over there. They will get in on the invasion I guess. They will really have the fun over there."

I don't know if he still had a glorified image of war or if he was just trying to send a message to Mamma and Daddy that like everything else, war was no big deal and he could take it in stride.

Curtis described his duties in a June letter,

"Boca Chica is a training area used by Ft. Brown on the Gulf of Mexico. Baker and I moved out here with the troops. While we were staying at Ft. Brown with the other instructors, we would have dry run classes of what we'd do out in the field. Baker and I taught the firing from the hip course, other instructors taught distance firing, others explosives, etc. There are nine different courses. There is an officer here, Col. Scott the Regiment logistics officer that I run around with quite a bit getting supplies that we need to teach the courses. He can get everything we need and fast. There's an officer and two men to each course. All these courses are to teach men to fight in the jungles. Every man in the Regiment will go through every course, and that will be about 1,350 men. We have just 10 days to teach them. I teach about 25-30 men at a time. Some of them are so dumb they don't know how to load their gun."

On July 25, 1943 Loyd was moved to Ibis, California to the Desert Training Center.

CHAPTER 11

THE TORNADO AND TRANSITION

TO MAINTAIN AN OFFENSIVE against Germany with the invasion of France a year away, the Allies invaded Sicily as the next step after Operation Torch in North Africa. The conquest of Sicily came relatively easy after some notable snafus during the invasion where our landing forces were attacked by our own warplanes. After conquering Sicily in July the Allies landed on the mainland of Italy in September to begin one of the bloodiest and most politically ridden campaigns of the war. Because of ego-driven, strategically unwise movements, the effort to drive the Germans out of Italy took months longer than needed.

In October, 1943 General Joseph Stilwell, with two Chinese divisions attacked the Japanese in Burma, near Ledo, India. He drove the Japanese back and at the same time began construction of the Ledo Road (also referred to as the Stilwell Road) which would eventually reach deep into the rugged mountains of Northern Burma and join the Burma Road.

On November 20, American forces under Admiral Nimitz invaded the islands of Makin and Tarawa in the Gilbert chain. Tarawa proved to be one of the bloodiest fights in the Pacific campaign. With bunkers carved into the coral, Japanese positions were impervious to naval bombardment and bombing. The island had to be taken one step at a time by the marines, using flamethrowers, grenades and rifles. We lost 991 Marines with 2,311 wounded. Only seventeen Japanese survived of the 20,000 plus on the island. (WWII, American Heritage)

Since we were being forced to move from the house at Lake Mineral Wells in the early summer of 1943, the year Herchel graduated from Garner High School, we had found another place to live. We had a crop in and had good, well-spaced rains and were looking forward to a bumper crop, but we had to leave most of that behind.

Finding a decent place was a miracle. "Miraculous' because the Camp Wolters training facility had brought thousands of workers, both civilian and military to the area, making housing a scarce commodity. We were in the process of preparing for the move and had devoted a day to cleaning and preparing the 'tile' house for occupancy. This was a departure for us, having lived in the less than sturdy houses of tenant farmers. We were no longer tenant farmers. We rented this house by the month. Still, it was Spartan in design and detail, a rectangular structure of four rooms of equal size. Two rooms were used for sleeping, one for a living room and one for a kitchen/dining room. The whole structure was encased by crude, cheap, red brick and tile produced from clay in the kilns of a company near Mineral Wells. We had a water supply piped in from an overhead metal water tank supplied by water from a well, pumped to the tank. We had running water in the kitchen, but still no indoor toilet. Actually this was a farmhouse, but the fields around lay fallow for two reasons; poor soil and an abundance of more lucrative jobs at Camp Wolters and the 'Bomber Plant' in Fort Worth, forty miles to the east. There was a barn and livestock shed that we used for a couple of milk cows.

On a day in late June, 1943 circumstances combined to bring us some very bad luck, touched with some extremely good luck. As we were spending the day at the tile house preparing it for our move, we couldn't ignore the storm clouds to the northwest. We knew we were getting another good rain on the crops we were abandoning. If Daddy had been with us he certainly would have declared the dark sky a storm 'that we would hear about.' By mid-afternoon we had finished our mission and had started back home to do some packing for the move.

As we emerged from the woods on the lane by the field approaching the house, all of us became uncomfortable by the eerie feeling of disturbance. Almost immediately, Mamma uttered a tearful cry. At that moment all of us recognized the reason for our unease.

"Oh Lord Jesus! A cyclone has hit. Have mercy. Have mercy," Mamma pleaded.

The heavily constructed storage building south of the house and our barn northwest of the house were missing. The sheet metal on the house roof was curled back. The watermelon vines were rolled and the corn stalks were askew. As we neared the site we could see that the house had been skidded back several feet, but was structurally sound. A tornado had struck. In a few minutes we had lost everything in the storage building and barn as well as the structures. We had canned hundreds of quarts of vegetables and fruit and had stacked them neatly in boxes in the barn. All that remained of this stash was a huge mound of fly ridden vegetables and fruit and broken glass dumped by a high-flying barn. Not a single jar was left intact. Evidently the tornado had spiraled the barn high into the air, dumping its contents into one pile before soaring to land and sink into the murky waters of the lake below.

A somewhat broad silver lining appeared, however. It wasn't long until we realized how great a tragedy it could have been. During the spring and summer, we had developed the habit of playing in and around the barn during the initial light sprinkles of thundershowers. As the rain would intensify we ran into the barn. Had we not been at the other house, we would have certainly been in the barn as it was destroyed. How many of us would have survived? Tragedy would be forestalled for now.

We didn't have a storm cellar at the Howard Place as we had had at our other places. Mamma would see to it that we were safely ensconced in one of these dugouts if a storm was brewing. She would proclaim that a 'cloud ' was coming, take the coal oil lantern and herd the younger kids into the cellar and agonize over Daddy and the older boys remaining in the house. In those cellars, we had the double stress of dealing with the storm and the ever- present spiders and fear of a copperhead dwelling amongst us.

With this event our move was accelerated and we began living in the tile house the next night and the complete move was accomplished over the next few days. We did most of the moving with our horse-drawn wagon and our Model A Ford sedan. On these trips we got as much as we could of the still plush crops. On the last trip in early July we just had to get the first ripe watermelons. Daddy prided himself in the growing of a very sweet yellow meat melon that he called the Bankhead. I know that our senses are dulled as we grow older, but I have never eaten a watermelon that could match the taste and texture of those grown by Daddy on that sandy loam soil of western Parker County. Some memories simply do not fade with time.

Jack wrote from Fort McIntosh, Laredo, Texas on July 8, 1943,

"Well I guess you are glad to be moved. Are you all going back to the old place and get any of your stuff? I mean the watermelons and garden, or are they moving the army in?"

We went back a couple of times to gather vegetables from the garden, roastin' ears from the corn field and watermelons that were just getting ripe. The tornado had stripped the plants and rolled the watermelons around, but they survived and the rain from the storm was enough to help make the crop. Of course, we got only a few of the watermelons and hoped that soldiers could enjoy the bulk of the crop. Most of the garden vegetables had been canned. Since the government had compensated us for the lost crops, we had no financial loss or complaints.

Apparently guilt was beginning to be an even greater factor in Jack's frustrations. His letter from Ft. McIntosh, July 11, 1943,

"I talked to Loyd today. He is doing o.k. He is fixing to go somewhere. He didn't say where. He said he'd write & tell me. I hope he gets to go out in Calif. I'd rather go out there as to La. But Louisiana wouldn't be bad without horses. Loyd wants to come back down here but I don't know why. If he can't come back I'm going to put in for overseas duty as soon as he goes over. I think I was the cause of him going to that school."

Loyd went to California, got to see the Rose Bowl game on January 1, 1944 and then on to Hawaii with his anti-tank unit. He would finish the war with occupation and mop-up troops on Peleliu in the Southwest Caroline Islands.

Jack was forever fretting about money. Another paragraph in the July 11 letter demonstrates the change in the family fortunes. Three years earlier the family income was near non-existent, except for The Boys meager jobs. Daddy was so broke that he borrowed eight dollars from Uncle H.D. to buy grocery staples. He wrote,

"Is Toughy and Jew still working? I guess they are making pleanty of money now. I guess you know we are all making about $700 per mo. now. If we don't save some of it we are fools. We'll see the day we'll look back & wonder why we didn't save some of it. I'm going to be able to save about a hundred dollars per mo. now that I've bought all the junk I need." [Jack had foregone his loan business for a time.]

Now that Herchel had graduated high school, he was eligible for the draft. Jack passed on some valuable advice to him in his letter July 16,

"Well Toughy I guess you are about to get into this little ragged, hungry army of ours. We just have steak ever day & more clothes than we know what to do with. I was just joking, but we have the best deal in the world I guess. The people at home are sacrificing a lot for the armed forces now.'

"I'd like to see you before you get into the army but I don't guess I can. I'll just have to write you what advice I can. I think I can help you a lot. ***Always keep your mind on your business. Be smart when you are dealing with non coms & officers. Always pretend to respect them wheather you do or not.*** Ask sensible questions but not too many. Always show interest & ***try***. Keep your equipment cleaner than anybody else in your outfit, shoes, rifle, belts, clothes & keep your lockers in perfect condition at all times. Take a bath every day in camp & keep your teeth clean & shave if you need to. Just be a good guy but don't let people get to you for money. Just say you are broke. When the others gripe just think how easy it is for you. If you want to you can keep that sheet & read it once in a while or I will send you a list of do's and don'ts if you want me to.'

He wrote of an unusual incident in his July 21 letter,

"A kid from G troop shot a horse today with a .22 and killed it. He was firing mounted pistol & started to reach up & get the reins with both hands & pulled the trigger & hit his horse right between the ears. Boy I'll bet he sees a big corner off his next 20 checks. These horses all cost from $160 up."

Another letter from Port Isabell on July 28 referred to the Holders' Chapel homecoming,

"I hope you all have a good time Sunday. I guess lots of people will miss it this year on account of gas rationing. I'll bet there will be pleanty to eat tho. I guess we will be back into Brownsville Sunday. I pity these guys riding back to Ringgold."

On other occasions Jack wrote about the usual stuff that is a part of our culture, as he did on August 8,

"Say I saw the best show I've seen since the war started today. It was, 'The Human Comedy' with Mickey Rooney & a little kid Jack Jinkens that's just about like Bill. It is one of the best shows I ever saw. When you are in town go up & ask Mrs. Dunn when it's coming to the Grand if it hasn't already & go <u>see</u> it."

And then he goes back to the main business at hand in his letter of August 18.

"Well it seems like the boys are going o.k. over in Sicily. They will probably take a few days breather before they start into Italy. I hope I get into some of it. There's a rumor going around that a bunch of officers are about to leave here. I might be one of them."

Curtis wrote August 24 about his and Joyce's move to a different house.

"I took off and we moved in a house to ourselves last week-end. It is a new house with 2 big rooms and a bath. It is just the right size for us, and it is a lot cooler than the other place. The best thing about it is that there isn't any dust. Joyce said she just has to sweep and mop once a day now. Billy, Mary and Lee are still where we were. We are paying $20.00 a month, but it is worth the difference. Jack came in the house the other day when we were gone, and Joyce found a note in the refrigerator in the empty bowl – "sure was good 'nanner puddin'."

From Fort Brown on September 7, Jack again wrote advice to Herchel,

"Toughy you just do what you think is best. You could put in for the Navy if you want to but you might not like it. Whatever you do get into just do your best at it. You can stay out of trouble easy if you half try. The army and navy are a lot alike when you are just starting out.'

"...I got a letter from Buzz today. He is doing o.k. He seems to be liking it o.k. He said it was hot as hell but he didn't even feel it. That's where that picking cotton & working in the hot weather comes in. We are used to it. They can't think up anything hard enough to hurt any country kid." I believed him.

On September 10 the U.S. Army invaded Italy at Salerno and became bogged down for weeks. Curtis wrote about it on September 12.

"The war in Europe sounds pretty good now. Probably sounds better than it really is. Italy is a long way from being ours. Heard from Buzz. He said they were really driving those tanks, but it's not as rough as the horse cavalry. We had dismounted and mounted review both. Saw Lee for a while. She is in love with an F Trooper named Mickey Crosland. He is a Sgt. from Palo Pinto. Been with us since 1940. I like him better than anybody in the troop."

Jack wrote September 19,

"Well what I didn't want to tell you was that I was taking an exam for promotion. I was afraid I would flunk it so I didn't tell you. I passed it o.k. & I will be promoted pretty soon but I don't know when. Don't address my letters to 1st. Lt. when I do get it because it's not supposed to be done."

He wrote again September 26,

"...We are all getting ready to go up the river. I'll be seeing Curtis & Joyce pretty soon I guess. We are going to take about a week going to Ringgold I think, and from there on North we'll all be together. We may end up in Santone about Christmas. We hear now that we will be back about Christmas. Boy I'm glad we are going out. I have it a lot better in the field. No barracks to clean or classes to hold. Just play war & take it easy. I'm glad I've had pleanty of experience living outdoors.'...

"I haven't heard anymore about my promotion. I guess I'll get it in about 1948. Heck I'm no 1st Lt. anyway.'...

"I'm trying to write & listen to Charlie McCarthy at the same time. I'm not doing much very good at either one."

He wrote October 3 with another update,

"...We are at Ft. Ringgold now. We got in yesterday. I guess we will leave about Tue. We are going on up the river. I think we are going on up to the Big Bend Country. We won't be back until Sprang I'll bet & maybe never. I don't care if we never come back. I'm like Buz said his C.O. is. He wrote Washington & asked if they remembered the 819th was still in the army...'

"Curtis is doing o.k. with his new job. He is dipping horses today. He will have to spend more time with the horses tho than he did with his men when he was on the line. His job now never stops.'

"Well we are going on field rations now. We won't be able to get all the stuff we have but it will be under army direction what we eat now & not just the idea of some mess Sgt. I think we will get along better. We will get a lot of canned goods, cured meat, and dried fruit. See we have been just buying our own stuff & getting whatever we want. I'll bet we have fresh meat when we get on up the country tho. They say the deer are too thick between Laredo and Del Rio.'

"How do you like your new job Dad" What kind of work are you doing? I guess it's pretty much difference in it & painting. Jobs around a job like that are usually pretty confining. Do you ever see Brooks Coalson? He should have a good job by now. He has been there every since they built that place hasn't he?'

"How are you kids liking Millsap School by now?"

Daddy had started work at the 'Bomber Plant' near Ft. Worth.

Jack wrote again on October 17,

"Oh yeah I got my promotion. Well I can use the extra money. I think it's $16 per mo. I'll be getting about $180 per mo. The order was dated the 8th of this month but I didn't get it till about the 12th."

Curtis wrote on October 23,

"Well I wonder when Herchel will settle down. At San Diego Calif., I guess. Buzz might get to see him out there. I guess Toughy was glad to get in the navy. I know I used to want in it but I guess I'm better off like this.'

"Well here comes the wagon train. The horse troops will be in pretty soon. Boy I wish you all could see this wagon train. They are all covered wagons & these old boys have those M1 rifles strapped on them just like in Indian days. This bunch calls them the '49ers. They have it pretty easy if the roads are dry."

This letter reflects the surreal combination of military units in this period of transition from ancient warfare to a more mechanized approach and the reluctance to give up one for the other. Jack and Curtis belonged to a unit that, with the exception of individual weapons, resembled closely the military units of the Middle Ages. Loyd was assigned to a mechanized cavalry unit and Herchel was in Navy boot camp preparing for duty on the U.S.S. Maryland, a battleship. This type of naval unit had reached its apex and was being replaced by the more effective aircraft carrier to destroy the enemy. This was just the beginning of a rapid transition to more efficient and damnable methods of killing our fellow human beings.

In September we began school in the Millsap School District. This was going to be hard to take, especially for R.C. who was a junior and had had to compete with the Millsappers all these years. Garner was smaller and always the underdog and in sports competition usually came out on the short end of the stick.

The Garner School main building housed grades three through high school. It was built of native sand stone by the W.P.A. The floor plan was simple with a small auditorium surrounded on the north, west and south sides by classrooms and the administrative office. Off the southeast corner of the building was constructed at ground level a huge stone water tank. Rainwater from the building roof ran into this tank and served as the school's water supply. At the base of this reservoir, hand controlled spigots were installed for students to use for drinking. We drank this water, richly flavored by decaying leaves and the juices of unfortunate insects that met their doom in its depths. Indoor toilet facilities or kitchens were non-existent.

The 'Dog House" contained the first and second grades and was located north and away from the main building. This was a white wooden building. An old school house from a nearby defunct school district was moved onto the grounds east of the Dog House to serve as a lunchroom as part of the recently created Federal School Lunch Program. I don't know how they managed without running water. Up the sloping playground to the east were the outdoor toilets, boys and girls separated a great distance, the dirt surfaced basketball court and the softball field for the elementary school. The high school softball field was further north down a lane near the railroad.

Interscholastic basketball was played on both dirt courts and in gymnasiums. The poor schools had to make it on dirt. The affluent, civilized school districts provided gyms. Whitt and Rock Tank, like Garner were dirt court folks.

The outdoor privies served a vital function, but not well. To begin with, despite the liberal use of lime, we couldn't escape the odor of their contents. As a result the teachers and principal would avoid them if they could hold it all day. This practice was not ignored by the older kids with the more seamy and mischievous intent. They would entertain themselves during recess from class by telling smutty yarns, smoking, practicing their cussing skills and other activities of a more vulgar nature. On one day at least the conversation turned rather deviant when one of the older boys rode his mare to school. This was my first exposure to those thoughts in relationship to animals. Boy! What would they think of next?

As I recall, my first encounter with real human violence occurred on these grounds. The last bell had rung to dismiss us from classes. As usual, most of us were awaiting the school buses to arrive for our trip home. An 'outsider', a grown man of twenty or so had been dating one of the prettier, popular senior girls. As girls will do, she also had a school boyfriend. The outsider didn't take this lightly, so he timed his arrival to coincide with our release. As predicted, his girlfriend left the building in cozy proximity to her school boy friend to walk to their respective buses. I don't remember any preliminaries but serious bloody blows were landed by both paramours. The younger man had an advantage in size and a considerable advantage in coordination over his drunken adversary. Notwithstanding, he gave up considerable blood from a lick to his nose. I doubt that either man won the heart of the lady, but I'm sure her ego was stroked by this battle of honor.

My first four years of school were at the Garner School. We had to walk to the bus stop, the distance varying from a few hundred yards to about a mile, depending on what farm we were working at the time. The first bus I rode was a green contraption that consisted of a plywood crate with windows and was attached to a chassis. The older kids sat around the perimeter of the crate on benches, their legs extended toward the center. The little kids sat astraddle a bench located down the center of the bus, boys on one end, girls on the other. School administrators are avid enforcers of anti-sexual contact and stimulation, not always successful. I know, for I later became one. I also know of the rate of success because I was also a practitioner of the illicit games we played as students.

We traveled the back roads of the school district to gather the school kids on the route. The main roads were gravel surfaced and provided reliable transport except in the wettest of times. The side roads were another story. On these rutted, mud-hole lavished trails the riders were frequently asked to get out and push the ramshackle old vehicle out of the mud holes or up the slick clay surfaced hills. This experience took the edge off the thrill of being late to school.

One of my pleasant experiences of riding a school bus was the peppermint stick give-away. An old farmer made a practice of riding the bus home from the store once a month, probably the day his pension check arrived. Each time he would have a bundle of peppermint sticks to share with the kids. These treats were much more treasured than the after-school snacks we usually got at home; a fried pie, teacakes, corn bread and buttermilk, biscuits and jelly…

I can't go on to Millsap without telling the story of the day I learned to never pre-judge a persons character. Everyone knew Clyde could whip ass. This was the persona that Clyde established by size, look and age. He was a year older than the other kids in his class and this gave him an automatic edge. Clyde simply looked pugnacious and he did nothing to discourage the image. He could get his way by simply saying it would be. All of us went along with this, even Rheudell.

Rheudell was slightly taller than Clyde, but his high-pitched voice and pigeon-toed, mincing walk gave him a prissy look, and he hung around the girls more than was acceptable by the pre-testosterone crowd. We never called him a sissy, but Clyde must have made the inference, a large mistake. Perhaps Rheudell had been coached by his he-man Dad, or by his no-nonsense Mom, but something triggered a livid and physical response. It wasn't your typical wrestling match with one kid eventually gaining the upper hand.

Rheudell went at Clyde with both fists at the end of his long arms delivering a devastating barrage of blows to Clyde's head and shoulders, giving him no semblance of a defense. He had no chance to mount a retaliatory offense. He was totally consumed with a futile defense. Clyde finally had to yell 'calf rope' and the fight was over. Rheudell may have been the class pacifist, but on that day he beat the hell out of the class bully. All of us were grateful to Rheudell, because we were able to live a relatively more peaceful life thereafter.

I also left some blood in the gravel playground of this site. I was rather slow afoot, but one day I was running downhill toward the Dog House, and gaining terrific speed. This new feeling was my focus, while the approaching tree should have been. I saw it and the kids playing in its shade too late. I had to choose the tree or the kids. My humanitarian nature took over and I chose the tree. The resulting trip to the principal's office for first aid gave me my first experience as a hero. I went back to class with a gauze bandage around my head to protect the wound. I felt mighty high, like one of our warriors in the battle against the Japs and Nazis.

So we gave this up with our move to Millsap, a school with indoor plumbing (One outhouse was available that the boys used for smoking and other clandestine activities.), gym, tennis courts, better buses and more kids. We joined the enemy. The three youngest, Roy, June and I, adjusted well and soon became entrenched Millsappers. However, R. C. couldn't make the transition. He started back to Garner School in the fall, transferring back to be with his old friends. He would never finish high school.

The family was in a holding pattern, knowing that eventually The Boys would go overseas and likely see combat. We would enjoy the good times as much as we could until that happened, notwithstanding Mamma's constant fretting.

The 124th Cavalry Regiment had acquired a new commander, Colonel 'Cactus Jack' Irving. He wasn't satisfied with activities around Fort Brown and the other forts under his authority. If he couldn't go to a combat zone, he would stage his own war with an extensive field exercise. There were some large ranches near Laredo that would serve his purposes of giving his men some much-needed training. It just happened to be in the heart of some of the best deer hunting territory in Texas. Jack wrote from there on October 31, 1943;

Mamma's forty-sixth birthday and my parent's twenty-eighth wedding anniversary.

"Dearest Mom,

"I hope you are feeling fine on your birthday. Just think…on your next birthday you will be Grannie Knight. You will really think you are smart then won't you. Well I'm sorry I can't help you out any on this grandma stuff. I guess Murphy & Joyce will make up for me & Buzz.'

"We are about 25 miles north of Laredo on the Santone Hwy. We are at the Callahan Ranch. It is on the Tex. map. It is a heck of a big ranch. The deer are as thick as rabbits. Don't look now but I shot one the other day right under the right eye. Boy we really had the steak the next day. I wish you all could have been here.'

"I didn't get to see Curtis but one time this week & that was Monday. I went over to see him today but he was gone. I think he had gone to Santone."

Jack wrote again November 4,

"We are on our way. I'm on the traffic control & go on ahead of the column. We are in Dryden now. This is one heck of a place. We went down to see Judge Roy Bean's place this morn. I sent you a purse yesterday, Mom."

On November 9 he wrote from Ft. D. A. Russell,

"I went up in the Davis Mts. Sunday. It is 22 miles from here to Ft. Davis & 17 more on up to the McDonald Observatory. That is really a sight and don't think the mountains aren't high. There was snow on top of some of the highest ones."

He didn't mention his frustrations in this letter but it accentuated his bizarre vacation-like assignment in time of war. Here he was visiting tourist attractions and going on a regimental deer hunt, justified as a military field exercise. It had to be eating at him.

CHAPTER 12

THE WAR BECOMES PERSONAL

ON DECEMBER 15, 1943, President Roosevelt and Winston Churchill met with Joseph Stalin in Teheran. Roosevelt and Churchill promised to open a front in France in the spring of 1944.

Men of the 124th Cavalry Regiment who had been transferred to the 112th Cavalry Regiment and transported to the South Pacific as part of the battle for New Britain were to meet death and injury. These men were the first of the original Troop F from Mineral Wells to face mortal combat.

The air war over Germany continued unabated with massive bombing by both British and American bombers delivering devastating destruction to German cities and factories. To reap revenge for the bombing of London, the British bombers concentrated on civilian targets at night in hopes of demoralizing the populace. Bombers from the U.S. Eighth Air Force were intent on hitting strategic targets such as factories and bridges and flew most of their missions during the day. The Allies were beginning to gain supremacy in the air with the introduction of fighter escorts that could stay with the bombers as far as Berlin. Hitler's mistakes in priorities led to a deficiency in well-trained pilots for the Luftwaffe coupled with a reduction in aircraft production. The Nazi air war was crippled.

Other action in the South Pacific continued through late 1943 and early 1944 with much of the fighting on and around New Guinea. The difficult Italian campaign was slowly progressing with much of the fighting from house to house. The rains, producing swollen rivers and knee-deep mud in the mountains and valleys of Central Italy, created impossible conditions for war.

By this time the Allies had lost thousands of men in heavy fighting on Salerno Beach and in New Guinea and the Solomon Islands. The impending invasions of the Gilbert Islands, Mariana Islands and further action in other areas of the Pacific Theater would take thousands more lives of U. S. soldiers, marines and sailors. Jack was taking all this in and was burning inside. (WWII, American Heritage)

Jack's lack of a direct contribution to combat was soothed minutely with his later assignment to advanced officer training back in Ft. Riley. He was being recognized as an excellent young officer and was being prepared for more leadership.

He wrote from Ft. D.A. Russell, Marfa, Texas on November 16, 1943,

"I don't know what I'd do about moving, but like you said you all couldn't hardly afford for Dad to quit work & you can't expect to do work in the field now. That is high rent for that place you are now but you might be doing better at that than you would otherwise. Maybe the war will be over in another couple of years & we can get a place to live for good. We want to save every cent we can now for the depression that will come later. I think you all could save more if Dad kept working. Because it's really hard to make $150 per mo. on a farm of any kind.'

"We are having a school now. I'm teaching in it. Three other officers & I are doing all the teaching. We have to go 8 hrs. per day, & prepare our classes at night. It will last for 15 days. I'm glad to get to do it because I can learn a lot."

Jack was very opinionated, even in the field of economics that, as far as I know, he never formally studied. His opinions were used to encourage the family to be conservative and head off the trials of bad times that many were predicting.

After he arrived at Ft. Riley, he wrote on November 28,

"Here I am back in this Cav. Center. I didn't know I was coming till the last minute & couldn't let you know when I was going to be in Dallas till it was too late. I sure would like to have seen you all but I guess I will live. I will be coming back thru in three mos. & I know I'll get a few days off then. They usually get around 5 days after finishing here. I was really lucky to get to come here. I'm the youngest officer here in grade. Most of them are from the early OCS classes & some are regular army & reserve officers. I don't give a dang tho. I can learn anything they can.'

"I think I'll go to the show this afternoon. Bob Hope is on."

Curtis, still on the 'deer hunt' writes on November 28,

"It sleeted and snowed on us all Friday night and all day yesterday coming in from the mountains. I killed a deer while we were out there. I slipped out of camp Friday about 4 P.M. with a rifle and 2 shells. Windy Andrews and Scroney were with me. We walked about 2 miles and jumped 4 big whitetail doe. We all three shot at them and killed 2 of them. We gutted them and went to camp and got a packhorse and went back after dark and packed them into camp. On the way to camp, it was sleeting so hard and was so dark we couldn't see our way. Our packhorse kept us from walking off a 50 ft. bluff by stopping and pulling me back just before I stepped over. We slipped into camp and packed the meat in our kitchen pack units (big metal pans) and told the Mess Sgt. to see that it got back to Ft. D.A. Russell with us the following day without anyone knowing about it. He did and I helped him cut it up into chicken fried steaks. The troop ate it thinking it was beef. Snake Warner the Company Commander

Joyce learning the proper salute

Herchel, the first to see combat

Daddy, Mamma, Bill, June at the Howard Place

Rio Grande City

complimented Doyle the Mess Sgt. He didn't know that it was venison, and never will."

Does fate intervene to fulfill a divine plan? You could think so in Jack's case.

He wrote the following on December 1.

"I just missed a trip overseas when I came up here. They had me on the order & then took me off. They sent 10 officers out of our outfit. Some of my best friends. One boy out of F Troop that I really liked. Curtis can tell you about him. He was an old Georgia boy. I sure would like to have gone along with them."

Our lives drifted along during the winter of 1943-44 with June in second grade, and Roy in the ninth grade. I was in the fifth grade, being introduced to classical music by our Yankee teacher, Mrs. Eter. Her husband was stationed at Camp Wolters. Herchel was on the Battleship Maryland in the South Pacific. R. C. was back in school at Garner. Daddy had gone to work in Ft. Worth at the Bomber Plant, building B-24s. He helped install the tail turrets and worked the swing shift. We seldom saw him awake except on weekends. We had certainly made some cultural adjustments for this war.

Curtis and Joyce were improving their life style. He wrote on January 1, 1944,

"Got our house on Ft. Ringgold. It has two bedrooms, bath, kitchen, dining room, living room and has a screened porch around most of it. It is on the north east side of the post."

For more than a year Daddy continued working at the Convair bomber factory. During winter days he harvested pecans on the Fort Wolters reservation. He knew the territory well. He had made a deal with the military authorities to gather the nuts on a 50-50 share. He and Mamma would work at this during the week and the kids would join them on Saturdays and holidays. Daddy would be instructed by the military where he could go to avoid the training exercises. We had a lot of fun playing in the foxholes and collecting empty ammunition shells and other interesting army trash. He knew what he was doing because we had done this before the war for farmers on the creek bottoms who needed help or lacked the inclination to do this difficult and dangerous work.

I had from an early age recognized that I had a bad case of acrophobia. I was then appropriately awed by Daddy's ability to climb and walk the limbs of those giant pecan trees that grew along the vicinity of the creeks that laced the cross-timbers area of North Central Texas. For the most part these were native trees that produced smaller, hard-shelled pecans. There were a few domestic varieties of paper shells mixed in but rarely had the quality or quantity of the native varieties.

Daddy would climb these trees that grew up to 100 feet and would walk along the branching limbs carrying a thirty-foot cane pole. He would reach up and out to thresh the pecans. I thought he was the bravest person I had ever seen. His skills became ever more awesome when, as an adult, I tried to emulate his performance. I found it impossible to handle a threshing pole with my right hand while grasping the nearest limb for dear life with my left arm.

The pecans would fall on the ground that we had previously raked clean, past the tree line. There the fun part started. All of us would crawl around on our knees to pick the nuts off the ground and fill our buckets. I loved the feel and smell of the pecans and the muffled, rattling sound they made when poured from our buckets into the tow sacks.

On these days lunch was memorable. We invariably had sliced white bread and lunchmeat sandwiches. I recall bologna, pickle loaf or liver loaf as the choice of meat. If we were lucky we would have potato chips. This was followed by an apple and vanilla wafers for dessert. We usually had water or iced-tea poured over ice in a fruit jar, sealed and wrapped in a damp cloth to keep it cool on hot days. We had warm days at times into December. For me this menu was a delightful departure from our usual fare of chicken, pork, fish, fresh or canned vegetables, buttermilk biscuits or cornbread, pies, cakes and cobbler. Anything store-bought created excitement undeserved.

After a days work we would take our sacks of pecans by a warehouse at the camp, weigh-in and divide the bounty. The income from this endeavor helped the family survive and even build a bank account for the first time since 1929. With my brothers safely tucked away on the border along the Rio Grande, at Ft. Riley, in Hawaii and aboard the USS Maryland, times were good.

My first awareness of a bank account and a checkbook came at this time. Roy wanted a bicycle, so we located one at a neighbor's whose children had outgrown it. They brought it to the house. Daddy took out his checkbook and very patiently and laboriously filled out the check and signed it. I was fascinated that he could get away with paying for something with a piece of paper that he had written on. I simply didn't understand why we hadn't been doing this all along. We wouldn't have had to work nearly as hard.

Jack wrote from Ft. Riley December 28,

"I heard from Herchel the other day. He said he really did have a good time at home & ate a lot. I guess he was anxious to get back, thinking he would get to sea. He might run into Buck out around the Hawaiian Islands some place. I think that's where Buck's ship is.'

"Dad how is your work coming along. I guess there's pleanty of work to be done & pleanty to learn around there. I wish I had a chance to learn something besides how to ride a horse. My great granddads 500 years ago knew as much as I do."

Still at Ft. Riley, Jack wrote January 23, 1944,

"I got a letter from Kenneth the other day. He is doing o.k. I wrote Bill Buchanan but never heard from him. I heard from Jack Stockstill too. His letter was written Jan 1, so I guess he wasn't shot in the New Briton invasion. They were the first to land. He said it wasn't fun like he thought it was going to be."

In a quick succession of letters beginning on January 31, 1944, Curtis told about the birth of his little girl.

"Glenda Joyce arrived 6:45 P.M. I helped Mrs. May and Dr. Ware deliver her, and when the nurse bathed her, I carried her to Joyce."

"This gal is doing perfect. She didn't give us any trouble last night. She's awake now looking like a little owl. She's been awake over an hour. We propped her up on a pillow and took her picture. She looked like she was smiling when we took it."

From Ft. Ringgold: "Got back here yesterday. Joyce will be here about Thursday or Friday. That will be a happy day. Glenda was getting fatter and prettier every day when I left."

Jack wrote January 30 with some worries.

"I wonder what has happened to Curtis. He hasn't written to me for a long time. I guess he's got something on his mind.'

"I sent Buzz a jacket today. He said he lost his coat. I guess he gets pretty chilly out there at night. I know how that desert is at night.'

"I heard Jack Langham was wounded in Italy & that Burgay was missing in New Britton. I wish you all would try & get some details for me on it, & let me know."

At this time Herchel had boarded the USS Maryland and was sailing by way of Maui Island to the South Pacific for his first combat experience. On January 31 the Maryland took part in the bombardment of the island of Roi in the Kwajalein Atoll, Marshall Islands. She stayed in this area until February 14 when an accident occurred on board. The primer from a 20 mm Jap shell exploded, severely injuring several men. She then steamed back to Pearl Harbor to place the injured men in a hospital and to make needed repairs. (Deck Logs, U.S.S. Maryland, World War II, Records of the Bureau of Naval Personnel (R.G. 24), National Archives at College Park, Md.)

Jack wrote on February 13, speaking of the men of the 112th Cavalry,

"That's too bad about the boys getting shot up. I guess a lot more will get it tho before this is all over. I wonder if Langham is back in the country or not. He's probably in a hospital somewhere. I guess Burgay just let one see him first."

"Looks like the invasion is coming pretty soon. They are really giving those Germans heck with that air power."

Nearing the end of his advanced training he wrote with frustration on February 17,

"Well if I knew all the junk they have thrown at us here, I'd be ready to go out & win the war all alone. I think they will have to add a couple of pack horses to the troops for the officers to carry their G.I. literature.'

"Well I gotta go to the graduation exercises. I guess they will tell us again, like they did at the end of OCS that we will be in combat in a very short time. I don't believe them anymore tho."

Even in the face of full knowledge that he could be killed in action, his frustration at not getting to fight was growing. He had a duty in this war and he wanted to get on with it. Another element of concern had entered the picture. Herchel, who had been in the Navy only a few months was headed for action in the Pacific. Jack didn't know that he was seeing action at that moment.

He wrote again on February 22,

"I'm o.k. but my morale is lower than a snakes bellie. I didn't know how it was to really gold brick. Well I'm getting a lot of it now.'

"When have you heard from that admiral? He must be winning the war all alone out there. I'll bet that little rat could tell us some lies now if we could see him. He hasn't written to me for over a month. I'll bet he was in on that Marshall Island battle. I don't think he is in any danger with the kind of Navy we've got. They are going to be out of anything to fight pretty soon if they keep on. They really let them have it at Truck but I don't think they had much there but aircraft carriers. I guess they had some protection but probably no battleships."

The USS Maryland had steamed from Pearl Harbor to Bremerton, Washington for dry dock and extensive repairs. (NACP, R.G. 243)

Jack returned to Ft. Brown and later wrote on March 20,

"I sent a package home today with some stuff in it for you all. I thought you all could use the ball glove & Dad that watch and brush might come in handy. Will you all keep those papers for me. Be sure & get two or three different ones on the day of the invasion & when Ger. quits & the Japs too. They will be good to look at later on don't you think. I'm going to send some more stuff home along & try to cut down on my junk around here so when I do leave I won't have so much to pack up & send. My crazy Capt. won't even talk to me about sending me off. He thinks I want to help win the battle of the Rio Grande. I think I'll shoot a Spick just for a little excitement."

On March 29 S1C Herchel Knight was given a twenty-one day leave. (NACP, R.G. 24)

Jack wrote April 6,

"I just got your letter today & about 30 min. later Joyce called me telling me that Toughy was there. Curtis had already called me & told me that he was home. I'm going up there Sat. morn. and I guess he will start back about Sun. evening. God I'm glad he's back & that he got into something. If I can't go I want him and Buzz to. I will tho before this is over. I think they are going to start some sort of a new Cav. outfit, & we are going to be part of it. I think. This outfit is going around in a daze & wandering what is going to happen. I hope we go to India."

Plans were in motion and the men were becoming suspicious. On April 14, Jack wrote, "Don't say I said so but I've got a suspicion that we may be going to Riley pretty soon. I don't know yet, but the rumors are really thick. The whole 124th may move up there."

Again on April 22:

"I saw Curtis & Joyce & the baby yesterday. Curtis came down with the Ringgold ball team to play this bunch. His team beat ours 6 to 1. Curtis got 3 hits for 3 times at bat. I guess he's about the best hitter in these parts. One of their horses won the open class of the horse show too. Lt. Pierson rode it. Curtis also ran the 220 yd. run & took second in it. We just had a field day with about everything you can think of.'

"I got a letter from Loyd finally. He is doing fine. I got a change of address card from him too.'

"Well I guess it's settled that we go to Riley. I don't know when tho. One of the officers here said he saw it on an order so I guess we'll go. I don't care. I'm ready for a change."

The USS Maryland sailed from Bremerton for the South Pacific to enter into the battle at Saipan. (NACP, R.G. 24)

Jack wrote again on April 30,

"We are all getting ready to move. I guess Curtis has told you all about it. We had been thinking we were going to leave for a long time, but this is it this time. We are supposed to be in Ft. Riley about the 12th of May."

"We are getting ready to leave. Our kitchen moved out of the post building on to the ground this morning. We have to clear the buildings 3 days ahead. I guess we will move out in Pup tents this afternoon or tomorrow.'

"Boy we had a mounted review yesterday morning that I sure wish you all could have seen. We had a retired regular army Gen. & a Col. out to watch it & they said it was the best they had ever seen. I guess we are pretty good.'

"The old General's wife is 77 years old & she rode a side saddle around with our new Col. while he inspected the troops. Her husband came along behind in a jeep. She is really a spry old lady. We have heard stories to the effect that she once rode with her husband & his Cav. outfit from Ft. Levinworth, Kans. to Ft. Bliss, Texas. I don't doubt it.'

..."I wish I could have been to the plays. I think I'll try to find out when they are going to have the plays in Junction City & go to them. I like to see kids like that. I'll never forget how scared I used to get. I know you kids did o.k. You are smarter than most kids. I think Mom & Pop kinda cheated some of us & gave to the others. I mean on the brains."

This may have been in reference to the Easter program. I got to stroll along the front of the stage in turn-of-the-century garb with Chrestine Young, the class beauty, and sing "Easter Parade." I fantasized about her being my girlfriend, along with most of the other boys in the class, through much of my days in school at Millsap. I think Dale Strain and Jack Parker had crushes to equal mine.

After he arrived at Ft. Riley, he wrote on May 16,

"We had a good trip up on the train. Capt. Mattson & I had a room with all the fixtures. We traveled like big shots. I just got another good look at Texas. Boy that is a big state & it's the best. These Yankees just stand with their mouths open at some of the sights you see in that state. All of us Texans make sure that they see everything that's good about it.'

..."There's sure a lot of artillery outfits here. We are in the 16th Corps now, but I don't know what part we will play with them yet. We will be going something tho if we go on maneuvers with them tho. We will probably do a lot of reconnaissance work. That means hunting the enemy & spotting them so the others can mow him down. That's what horse cav. does mostly. We are going to get a new shoulder patch. We have pulled off our old ones already."

Jack had moved into a new world of the semi-privileged officer ranks. He appeared to be incredulous that a tenant farming country boy could find himself in those circumstances.

He wrote again on May 22,

"I'm writing in class now. We are having classes on foreign maps. We will study France, Germany, Holland, China, Burma & India. We just got hold of a lot of maps & decided it would be a good idea to study them a little."

Curtis wrote May 23 from Ft. Riley,

"Joyce and Glenda are here now. We have an upstairs apartment in Manhattan. It is nice. Mary and Sandra are here too. They are staying in a room in a hotel." Sandra was the daughter of Billy Ralph Aaron and Mary.

CHAPTER 13

THE BOYS GO OVERSEAS

IT IS BLASPHEMY to mention Jack and Joseph Stalin in the same paragraph. Without a doubt, however, they were experiencing the same emotions in June, 1944. Stalin had been insisting since late 1941 that Churchill and Roosevelt open a second front in France to ease the pressure on the Soviet forces in the east. He was to finally get his wish on June 6 as the combined Allied forces of the United States, Britain and Canada successfully pulled off history's greatest amphibious landing on the Normandy coast of France. Stalin was elated. (WWII, American Heritage)

Jack was at Fort Riley, Kansas processing for his long anticipated and greatly desired overseas assignment. He didn't know where the 124th was going, but it mattered little to him. His preference would be to get a chance at the hated Japs, but Germans would do. He was satisfied at the thought of finally fulfilling his longing for action and a fitting culmination of his duty and ultimate destiny.

At the same time our forces in the South Pacific were increasingly pounding the hell out of the Japanese. By June 10, the USS Maryland was anchored at Roi Island, Kwajalein Atoll, Marshall Islands to prepare for their next venture. One week later the ship was lying off Saipan delivering ammunition to the USS Louisville and positioning for the bombardment of the island. The Maryland and her sister ships were extremely busy the next several days with their bombardment duties executed amidst fighting off enemy aircraft and the Japanese navy. (NACP, R.G. 24)

On June 15, 1944 U. S. forces invaded Saipan with the usual bombing and bombardment and the usual air fight with the Japanese carrier based planes. Four days after the invasion, the U.S. and Japanese engaged in a large and intense carrier battle. It was an overwhelming victory for the United States forces. Shortly thereafter, on June 22, Herchel's ship, the USS Maryland was hit. The deck log made the following notation:

"1952 A Mitsubishi Type Dive Bomber appeared over the bow of the USS PENNSYLVANIA on bearing of about 030 degrees T., 30 degrees off our bow, distance about 1200 yards, headed for the ship and banked passing down the port side of the ship, afterwards the ship was struck by a torpedo between frames 8 and 11."

The deck log also has this entry made July 1, 1944,

"The Maryland's torpedo damage is as follows as observed by COMSERON-10. Aircraft torpedo hit port side at frame 8 about second platform level. Plating and structure port and starboard blown away between hold and third decks and between frames 4 and 10. Damage to plating extends aft to frame 13. Starboard plating blown clear of framing up to second deck. Wrinkles in plateing port and starboard extends to forecastle deck at frame 8. Keel structure in place but badly distorted forward frame 12. Stem casting broken in 3 places between third deck and forefoot and bent to starboard. Bulkhead 15 between hold and first platform undamaged. Bulkhead 14 between first platform and main deck undamaged. Condition bulkhead 14 between A1V and A2V unknown but A2V flooded. Bulkhead 11 between main forecastle decks undamaged. Bulkhead 9 between second and forecastle decks badly distorted. Second deck undamaged aft frame 14. Second deck between frames 9 and 14 badly distorted but in place. Second deck between frames 5 and 9 badly distorted and non-watertight. Main deck aft frame 11 undamaged but badly distorted between frames 5 and 11. Forecastle deck and fittings undamaged but deck has indication of slight droop forward frame 9. Section of bow forward frame 5 and above second deck undamaged. Salvage this section considered possible. Hatches in first platform and third decks in WT trunk frame, 13 blown off and trunk boundaries slightly distorted. A-304-A flooded around spring door leading into above trunk. All space forward frame 14 and below second deck open to sea. New bow structure required forward frame 14 from keel to main deck and forward frame 11 above main deck. All electrical and mechanical fittings included in new structure require replacement."

The Maryland was returned to Pearl Harbor for repairs. Working from construction plans of the Maryland, the workers built the replacement bow while the ship was en route. The ship anchored at Pearl Harbor on July 13 and by August 13 she was on her way to Palau Island for bombardment duty before the invasion of that island in mid-September. (NACP, R.G. 24)

The 124th Cavalry was in a state of feverish preparation for its next move. Jack wrote on June 7,

"Well the big show, as they call it, is on. I sure would like to be there. I just wonder what it's like. They seem to be doing o.k.'

"Well we are getting ready fast. We are checking all our clothes & are firing all kinds of weapons. We should be taking off in a couple of weeks."

The Fifth Army had entered Rome on June 4 with General Mark Clark touring the place the next day, elated by the fact that he had accomplished this feat the day before Operation Overlord began. (W. W. II, American Heritage)

Jack wrote again on June 15,

"This dang troop is just like a mad house today. We were working all day in the orderly room here & just in the next room an old bitch has four pups. When it got hot today they started howling & kept it up all afternoon. Capt. Matteson nearly blew his top & started to kill them. They even got on my nerves along with about a million other harassing agents. We will have our troop on the line in a few days tho. I wonder how Snake & General Whatley are doing. I'll bet they are having hell too."

He was making reference to Lt. Jack 'Snake' Warner and 1st Sgt. Sam Whatley, who were F Troop Commander and First Sergeant. Jack was a serious person, devoted to duty, but he also had a sharp, abrasive sense of humor that could turn self-deprecating.

In the last paragraph, he continued, "I got my income tax return money that you helped me figure out Nig. It was $63 & I was sure surprised. I never expected it. I wonder if I'm still getting my $50 bond every mo. I'm going to make a pretty good allotment when I start to leave. Had you all rather I sent it in Bonds or cash. Curtis can tell you what we said about the place."

Apparently Jack had thought they would be going to Europe, since horse cavalry would be more suitable in that terrain than in the jungles of the Pacific Islands and Southeast Asia. But the military strategists had conceived another use for these horsemen.

Curtis wrote on June 24,

"There are rumors floating around that we will be going overseas pretty quick. We've been hearing that since before we left Ringgold. We hear that we may be joining the Merrill's Marauders in Burma." Unknown to them, this is one time that the rumors were correct.

Their letters kept coming. Jack wrote June 28,

"I haven't heard from Loyd or Toughy for quiet awhile. I know nearly that Bape has been in that last big fight. I guess Buck & Glen McCreary were there too. Glen is on a carrier now, & they had about a thousand carrier-based planes. I sure would like to have seen that outfit when it took off."

With the impending departure of the 124th Cavalry to a destination unknown to the troops, there was much speculation about the direction they would go. Jack was trying to guess about where Herchel might be July 4.

"I can't tell you what or when we'll be doing anything because it's against regulations but I don't know anyway & don't want to know. I'm glad of one thing that I've learned in the army so far is to obey without question. I don't even consider wheather it's right or not...I just do it....'

"I sent some more bonds this mo. I can't think of what Herchel was talking about unless it was that they got a crack at the whole Jap fleet. I said I hoped they did, & that they had to hunt for Japs."

At last they were moving from Ft. Riley on the journey overseas.

July 10, 1944, APO N.Y., N.Y.

"Well we are here! At last we seem to be on our way somewhere. I can't tell you where because I don't know & I wouldn't if I did. Don't try to guess & be telling people because you might be right!"

The 124th Cavalry had been moved by train to Camp Anza, south of Los Angeles. They were there a few days waiting for their equipment to be loaded and last minute processing to be finished. The regiment boarded the USS General H. W. Butner and left port on July 25, 1944. The New York APO address was to confuse the enemy as much as possible. It probably only confused the troopers' families and friends for a short time. (Marsmen in Burma)

Loyd wrote to Jack from Hawaii on July 13, 1944,

"Got a letter from you yesterday. Certainly was glad to hear from you.

Guess you all are on your way to parts unknown by now. You said in your last letter that you couldn't say what you were doing. I have a pretty damn good idea. If it is what I think you probably had some good headaches.'

"I suppose your right about us having a place in this war but I can't figure it out yet. We're just marking time here.'

"Well I've got to go to a softball game. I may 'twirl' a few innings. We've got a pretty good team. Haven't lost but one game in the last half of play."

Jack wrote July 16 from Camp Anza,

"...I haven't had a letter since I left Ft. Riley, but I can't say anything because I haven't written but one. I'm writing Herchel & Loyd. Herchel said he was in on the Saipan deal. He's an old veteran by now."

When The Boys wrote each other their language became much more colorful with shades of blue. Loyd wrote Jack on July 17,

"I don't know where you are by now but I imajine you are takeing off for some unknown place. I hope your satisfied with where you land. It has took the 124th a long time to get started but I suppose they will be plenty rugged when they do get a chance to do their stuff. Wish we were all back in F Troop like we were when we first came into the army. That was a good bunch of men, although nobody realized it at the time. Theres been a lot of them made officers since Nov. 18, 1940.'

"I had just started this letter when the mail orderly came in with the mail. I got a letter from Mama in which she gave me your address. Boy I hope you guys don't get into anything hot over there. I was hopeing you would come this way but I suppose they need you over there worse. Do you think you will ever get your horses back? It doesn't look that way does it?'

"Things here are getting mighty stale. We are still marking time here on Hawaii. I guess we're lucky and don't know it. Everybody bitching all the time. The morale of this outfit isn't too hot at the present.'

"One of my best PFCs was sent to OCS at Benning today. I'm gradually losing my best men. It wont be long now until I will have an outfit of Reime's. Not that bad I guess but bad enough that I wouldn't wont to go into battle with them unless I had about three months of rugged training first. I still have my old NCOs tho and it won't take long to get another bunch in shape.'

"I guess Mama wrote you that Freddy got married. That takes the cake. If he can find one that will have him I think I should try to find me one. She also wrote that Tuffy Holder was married. Seventeen is kinda young but I guess he could see that draft notice comeing.'

"How is our duck legged friend comeing along. I'll bet the little bastard is scared to death. Tell him that I sure hope he doesn't get so damn sea-sick that he falls overboard in the most shark infested part of all oceans. I'd like to see him with one of those V-1 robot bombs nipping him in the ass.'

"Well tell every one to take it easy on the Jerrys. I'll be writing again soon."

Mamma wrote July 17,

"Dearest Jack,

"How are you by now? We are all well as usual-hope you are ok.'

"It sure is hot here now-been hotter since the rain than before.'

"We had a letter from Curtis this AM. Was glad to hear from you all-he said he seen you ever day.'

"Joyce & Glenda left yest. for Poteet. We sure hated for them to leave they were so much company. Glenda is so sweet & getting smarter ever day. She will stick her lips out with her mouth open & act so cute.'

"We had a letter from Loyd & Herchel yest. Herchels was written the 12th of July. Just took five days to get it. Loyds was V Mail & took 11 days. They were ok. I think Herchel was at Hiawaii.'

"Dad & kids went to Garner this eve. Wade was up there he said Rowe was fixing to go across.'

"Dad went to see the Dr. this morn- & he wasn't there. He was to go back today to see if or when he could go to work.'

"Jack if you have a receipt showing you expressed your things, if you would send it. We could get them to see about your things. Or you could write them. They told Dady to write the express office, so he may write up there. Joyce was nearly a week getting her suit case.'

"Aunt Maedell & Lorene came by & stayed a little while yest. Said Mike had been taking some hard jungle training. He had been wanting Loyds address so he could find him or try to. Mike he is cpl. I'll send you Mikes address.'

..."His A.P.O is 957 & Loyds 960. Reckon they are close together.'

"That explosion sure was bad in Calif. wasn't it. I couldn't keep from wondering about you and Curtis, tho your address is New York. Well I can't think of any more to write so will close for now."

"Write real soon & be sweet & careful. All our love,

Mother & all

Jack wrote again from Camp Anza, July 20,

"Well here am I, with not a worry in the world. I'm in the best humor that I've been in for about four years.'

"Well the war news sure looks good. I just made a $20 bet that Ger. wouldn't be out of the war by Sept. 1st. I hope I lose. If I lose I've still won. I know that's against your belief Mom but I'd give anything if it would stop. I still want to go over tho & see what it's like."

Jack's ambivalence shows. After years of guilt and disappointment, he was finally on his way to fight the hated Japanese. On the other hand, he wanted no more killing and he had a normal sense of self-preservation.

In a faraway land a military unit was being formed. On July 26, 1944, the 5332nd Brigade was formed. In another month the 124th would show up to take its place in this force. (Marsmen in Burma, John Randolph, Gulf Publishing Co., and Curators of the University of Missouri, 1946, 1977.)

On July 21 Jack wrote again from Camp Anza,

"I'm actually getting low on money for the first time since I came into the army. Those gals around here really get the money. I love 'em all tho.'

"Curtis is taking it easy. He stays in every night. Well I guess if I had an old lady & brat at home I'd be staying in too, but until I do have, I'll keep up all the feminine morale I can. I like to make everybody happy."

"We had a good ball game yesterday. The anti-tank Plat. played the second Plat. & beat them about 6-0. I umpired & they nearly killed me.'

"Well the European situation is certainly looking a lot better every day. I think, it's about to come to a head over there. Let's hope so anyway.'

He was certainly teasing Mamma. He knew about her Victorian views, but he had acquired a state of euphoria about his chance to fight and his light-heartedness showed. They would board the USS Butner and sail from Los Angeles Port of Embarkation on the morning of July 26, 1944. There was no more communication from them until they reached their destination. Excerpts from several letters will tell the story of this period.

Loyd had written Curtis on July 18,

"I'll bet you guys are having a hell of a time right now. Of course I know you'll miss Joyce & Glenda.'

"It'll probably be quite a while before you get this letter, that is if you've taken off before it reaches you. I wish you had of come this way. Maybe you'd get stuck here on this island and we could have gone on to do a little combatin'.'

"Everything here is still static. We have no cause to be thinking of moving anytime soon so I guess I'll be a veteran without any wounds except those I receive in brawls or jeep wrecks."

Mamma wrote Jack on July 21. I don't know if it reached him before they sailed. Excerpts follow.

"The kids were glad to get the money. Bill & Jr. say they are going to get them some school pants. Rodeos I think.'

"We have heard from all you boys this week. Had 2 from Loyd & Herchel. They were ok. I think they are pretty close together. When we hear from one we hear from the other.'

"We had a card from Mary this AM. She said Joyce didn't have time that morn. & would write that night. Said Curtis called the day before & was ok.'

"We have had some hot weather but it's cool yest. & today. Been raining today.'

..."What do you think of the Japanese shake up. Seems as if they & the Germans are not doing so well are they. I guess they are getting nervous. I hope it all turns out for our benefit. Mabe this war will end sooner than we thought.'

"We've been attending the democratic convention, at Chicago right here. They sure have it some time. Seems Texas isn't for Rosevelt very strong, but I won't be soprised if he is elected. He was appointed canadate for President again. It might be best I don't know.'

"Wanda and her husband & 3 kids came by. They had been to see Aunt Nell. Wanda looks kinda old. She was laughing about you & her eating a race to see which one could eat the most biscuits. I told her I bet you started it so you could get all the biscuits you could eat. Aunt Nells biscuits were so small. Wilfords still at Lubbock. Inez is in Philidilphi with her husband.'

"They said we could realy make the money out west. Said Jr. could get $5 a day, R. C. $1.00 per hr. I would like to go when Dad gets able to work, or get through with his treatments. I think it will be soon. He's lots better. June wants to write, so by & be sweet."

Love always, Mom

June added her letter,

"What are you doing now. It rained today and last night. Wanda and falmy came today. She has two little girls about my age, one 6 years old and one 10 years old. R.C. is up at Sims now. He has been going up their 3 nights.'

"Thank you for the money you sent me.'

I love you,

Love June

Mamma wrote Jack July 24,

"I sure hope you loose your $20. As you said you would win if you did loose.'

"We got your pictures & other things you sent in the letter. Your picture is good.'

"We had a picture of Glenda made with a Kodak, so I am having some made to send you boys, it is so cute. In her swing with a little print sun dress on. She looks so big in it.'

"Did I tell you I heard from Aunt Bessie. She said they like Calif. & wanted us to move out there. Said they still wear their coats & sleep under blankets. Said Evelen was making $27 per week, but I know we never would get that far.'

"We went to Howard's yesterday. They were about as usual. They said they would come down here one day this week & stay a few days if we would go after them. So I will wash & clean up a little first."

Mamma wrote Curtis July 25,

"I hope you are still feeling ok. & still gaining & get big & fat. We are all able to eat all we can get, also work. Vercia Lee & I did a big washing this AM.'

"R.C. started out west this AM to visit the Stubbs & work a while. When he was getting ready to go we had a letter from Herchel telling him to go out there to see the Stubbs. It was funny.'

"We had a letter from Loyd & Herchel too. They were ok. & I think both in Hiawaii. I wish they could get to see each other. There is several boys we know over there. I wish they could all get together.'

"We had a letter from Joyce last weekend. She said they were ok. I sure would like for them to be up here, but I know she wants to stay most of the time with her mother, tho she said she would like to stay here, but she wanted to go be with Lee.'...

"Norma Sims is here and talking till I can't think of anything to write and there's not much news either.'

"The war news sounds good. Looks like the Germans are going to have to do something soon. The sooner the better with me.'

"We went to Howards' Sun. I guess we will go after them about tomorrow...They sure have a pretty crop. Makes me want to farm. I thought that old place was worn out, but I guess they could make a good crop on any kind of land. They loaded our car with watermelons, tomatoes, potatoes....,

"I look for Dads work to stop anytime. He isn't working now as he had to take more treatments. I'll be glad when they get through with him or when he gets well. They should have cut them [hemorrhoids] out and get it over with, but he feels ok & eats as much as ever."

Mamma wrote Jack again on July 28,

"Dad took his last treatment this A.M they said, so I'm glad. He was feeling pretty good before he went back, but they said there was one little place needed treatment, so I think he will be rairing to go to work soon....'

"I saw Lorene & baby at W'ford this AM. & her sis in law & her little girl. Lorene said tell you she sure likes her hose. Said she had worn them a dozen times. She wore them to a wedding....Said she almost held her breath afraid she would tear them.'

"If I don't forget I'll send you one of Glenda's pictures. I think she is so sweet.'...

"Josie & Wade are here. Dad went after them Wed eve. I think they want to go back this eve. I wish they would stay till Sun. they are ok. Rowe is still at the same place, or was the last they heard. Connie still working.'

"R.C. went or started to West Texas Tuesday. We had a card from him yest. He was at Olney Tue nite. Said if he seen some of Bryants folks he might stay there & get an early start next morn. He was headed for Stubbs at Floydada. They are paying high wages out there, but I look for him back in a few days."

A letter from Mamma to Curtis July 30,

"I heard from all you boys this last week & I'm thankful. Also heard from Joyce twice. They were fine. We can't hardly wait till they can come back up here.'

"I do hope you & Jack don't have to go across. I feel like the war might be over by Xmas. I believe the Japs will quit soon after the Germans are whipped & don't look like they will last very long now.'

"Curtis, R... G... was killed in a car accident Sat. morn. about 2 oclock over here this side of the under pass just over the 2nd hill from here. There was 3 women with him. A woman was driving. It was her car & it killed her. They were in the front seat. They passed a car going 55 mi & after they passed it they run the car up under a Merchant truck. It broke both their necks. You know he was married & his wife has a little girl. That is bad to be killed in that condition but sometime I think the world would be better off if all men & women like that was out of the way. Anyone that has a true companion should be thankful. I'm glad that Dady has always been decent.'...

"I forgot to tell you we didn't know the woman that was killed. Lived in one of those cabins up at the station. She drove a truck in the camp & was a soldiers wife. They said he came up after it happened. I guess they went after him & he picked her hat up & said just as I expected. You are as dead as a mackeral. So I don't suppose it worried him very much.'...

"I saw Anna Coalson at W'ford yest. She said Kenneth was starting for home Monday on his first furlough. She was so excited. She began to tell me how hard it was, like I didn't know. I know our Lord has helped me & I'm so thankful that he has protected you boys."

Mamma continued the stream of news from home. Her letter of August 1,

"Dad has been sick since Fri. but seem to be better this AM. I sure don't want him to take any more treatments. It is worse than the disease. He should be o.k. after this one.'

"Uncle Bill came this AM. He's going to Weatherford to see about getting a place for V. Lee. All the kids has refused to keep her so it isn't fair for one or two to have to keep her and they would try & put her off on me but it won't work. He wants me to meet him in W'ford this eve. He went on to Mr. Hills.'

"I sure hope you & Curtis are on this side. I wish you and him could get a furlough & come home. It would help dady more than any thing & me too, but if you boys can take it we can too. So don't worry, we'll keep things going till you come home.'

"We rec. your letter Sat that was written the 21st. Was glad to hear from you. I thought about sending you some money but you would have got your check before you could have got it. Howards were here & we laughed about you spending your money on the gals. I told them I bet you didn't hurt yourself.'

"Honey we still haven't gotten your things. I think I will write up there & see if I can do any good. You must not be getting my letters or you would say something about your things."

I was gaining small amounts of insight into the man and woman thing from my brothers. If I had waited for Mamma and Daddy to tell me, I would have remained ignorant, not just misinformed. One of the greatest surprises came in the summer of 1944. We had found a good, deep, clear swimming hole in the creek behind our house. It was on the property of our neighbors who had daughters the age of my brothers at home. Somehow the word got back to us that these girls had hidden in the bushes and watched us skinny-dipping. In retrospect I know they weren't interested in my pre-pubescent assets, but my adolescent brothers'. The surprise at the time was that girls were interested at all.

Mamma wrote Curtis August 3,

"I had a letter from Joyce yest. telling us the news about the letter she got. I sure hope you all arrived OK & are safe at all times. I feel like you will be, tho I'd rather all of you be in the USA. I hope you & Jack get to stay together.'

"R. C. is home. He stopped & stayed last week with Aunt Doll & worked 1 ½ days so aunt Doll sent Vercia a card & said he was there, so I phoned for him to come home. I didn't want him up there with Bryant. He'd have him honky tonking.'

"H.D. said Bryant got him started tho he ought to have better sense now.'

"Uncle Bill & Cush came Tue AM. & stayed till Wed AM. He is trying to get a place at Austin for V. Lee. Mr. Hill is helping him. They will find out in a few days. They say it's a nice place & they will teach her to read & write if she can learn & I think she can. She is anxious to go so she can have a room. I hope she can have a room to herself & be satisfied but I know she will worry for a while.'

"Curtis, Mrs. Maisures (Mrs. Andrew) was buried Monday. Wasn't sick very long. Something wrong with her vaines in the legs."

Curtis had worked in Mr. Measures peach orchard the summer of 1940.

Vercia could learn. I taught her to read some during a couple of sessions at this time that she lived with us. She also initiated the solution to another problem that had arisen.

Even though we had given up on growing crops, we still maintained vestiges of a farmer's life-style. We kept a mule and horse in reserve for the future. The milk cows continued to provide a ready and ample supply of dairy products. What is a farm without the chickens? We needed eggs and the occasional fryer for Sunday dinner.

A rooster was always a part of the flock, either from habit or from the misconception that contented hens produced more eggs. This tradition helped create the bad ass rooster episode.

We kept White Leghorn chickens. Our rooster at the time grew into a giant for this variety of the species. For reasons never quite defined, this anomaly also became very territorial and hostile toward the bipeds living in the tile house.

He staked out an area between the house and the outhouse and barn as his sphere of influence. He took great exception to our infringement on his chosen ground. Now, this was a smart old bird. He left Mamma and Daddy alone, but he would attack, with great vigor, June and me as we courageously ventured to either of these outdoor facilities. He made one mistake. Apparently he mistook Vercia Lee, Mamma's mentally retarded sister, as an oversized child. He didn't account for Vercia's having grown up with a bunch of baseball playing brothers. As he launched his attack, Vercia grabbed the closest stick she could find and, with uncanny aim, whacked the old white rooster up side the head.

I didn't see the lick, but I heard the disturbance and caught the drunken dance of this tyrant of the barnyard and heard Vercia's colorful and somewhat vulgar tirade directed at her attacker. This was enough for Mamma, and with no attempt to revive the distraught rascal, she finished the job by taking up the hapless bird, wringing off his head and vengefully transforming him into a delicious pot of chicken 'n dumplings. Nobody or nothing messed with her kids or her childlike sister.

Loyd wrote to Jack on August 3,

"Dear John,

"Haven't heard from you or Murphy in over a week. I'm just wondering where in the hell you're at now. I wish I could be along wherever it is.'

"This is a darn good place if you like to go to bed at nine o'clock every night. The officers here are restricted from going to just about every place in town. There's only about four places that we can go to drink and raise Hell. They all close at 1900 every day except about twice a month they stay open until 2100. So theres not a hell of a lot to do except go to a show or stay home. I have an ice box in my hut that I keep beer & cokes in. We use the cokes to mix drinks. I'm getting to be a regular damn sissy, taking chasers with my liquor. The "all stars" and the rest of the F Troop "regulars" would disown me if they caught me at it.'

"Our softball team lost a close one last night, 3-2. They were playing the leading team. They have won fifteen straight games without losing a one. I guess I'll have to limber up and twirl a game against them. I guess we will start playing basketball in about another month. I think I'll try a little of that.'

"From all indications the Russians are about to win the war in Germany. Maybe it won't take too long after that to finish this side of the fracus."

Herchel wrote a letter to Jack one week before the Maryland sailed from Pearl Harbor, heading for the South Pacific. Jack and Curtis were somewhere on the same ocean. It was August 7, 1944. The U.S.S. Butner had crossed the equator August 2.

"Dear Jack,'

"I have a little time so I'll write. I haven't heard from you in quiet a while. I guess you are a little busy.'

"How is Murphy? Does he miss Joyce much?'

"I got a letter from home about three days ago and Smutter is up at the Stubbs kids house. Kenneth Stubbs is in the Navy now. Just going through boots. I feel for him.'

"I guess Mom told you about R… G… being killed in a car wreck. I hated to hear that. I guess if he was where he should have been ever since the war started, it probably wouldn't have happened.'

"I sure hope you guys come over this way. We all will probably be in on the next one. The next to the largest I hope.'

"I guess Loyd is still over here. I haven't heard from him for a spell. He saw Bob Hope, or had a chance to, but didn't go.'

"Well John I'll close for a while.'

"Take it easy and let me hear from you soon."

Love,

Herchel N. Knight

Jack had mentioned his girlfriend in his letters. Apparently, they had kept up correspondence since he left the border. We have one of her letters that survived the decades. She wrote August 10, 1944 by V Mail,

"Jack Darling, have been worried about [you] because I haven't had a word from you in a long time. Hope that you are all right and that you will try and make time to write to me, dearest. Wrote to you about a month ago – hope you got that letter.'

"Just think love, a year ago, you were here in good old Rio, and I was seeing you as often as I could. Just yesterday, Chichi and I were thinking of the time we had gone to Fort Ringgold and you told Chacha what pretty legs she had – remember, love? Also about that time that Chichi left the show with Lt. Johnson and her mother came in looking for her and I was so afraid. Those were good times for me, but I doubt it, if they were for you.'

"I went to Alex. Louisiana to visit my brother. He is stationed at the Air Base there. We had such a good time together. Dad heard from him yesterday and he said that he sure misses me now. I stayed with him two weeks.'

"Well darling, I better sign off now. Please love, take care of yourself and write soon. Think of me often – I always think of you, dearest.'

"All my love,

Maudie"

Mamma wrote Jack on August 12,

"Dearest Jack,

… "We are doing fine. Dad is up eating all he can get & helping do the house work.'

"I write & keep writing. I don't know when you will get it, but when you do you will get all at once I hope.'

"Howards came & spent the night Thurs night. They are ok. Rowe sent his sleeping outfit home so I guess he is going over.'

"We fixed to go home with them yest AM. But little Bobbie fixed up to go too. So we didn't have room for all of us and we decided not to go & he didn't offer to go home. Don't he need a hitch in the army to learn him something. He quit the brick yard because Garland told him to pep up.'

"Uncle Henry & Bobbie spent the night with us last night.'

"We heard from Herchel this week is only one we have heard from. He was still at the same place.'

"Jack for fear you haven't got my other letters, I'll tell you again. We haven't got your things. If you have the receipt, send it & they said they would trace them for us, or did you turn them over to the well fare Dept. If so they said it would be a long time before we get them.'

"Are you & Curtis still together. I sure hope so & hope you are both well.'

"I'm in a hurry it almost mail time.'

"Write soon as possible. All our love & God bless & keep you always.'

Mother & all."

Excerpts from Mamma's letter to Jack, August 16,

"It don't seem right to write & not hear from you boys. It don't seem like you will get my letter, but mabe you will after so long. I do hope so.'

"There isn't much news to write. I wrote Curtis all the news. So I hope you & he are together.'

"I had a letter from Joyce today. They were ok. Glenda is crawling and we had a letter from Herchel today and one from him & Loyd Monday. They were ok. Herchel sent 2 little pictures. He looks pretty good. He sent the kids $10. Sent June $5 & Bill & Jr. $5 between.'

"Loyd sent a $50 bond. Said he might change his allotment to bonds.'

"R.C. went to work at Camp Wolters in the service club yesterday. Suppose to get over $100 per mo. Jr. has been hoeing for McKinsy & he wants one of them to help make syrup the last of the week or next.'

"I wrote Curtis about Aunt Ella coming down. She came on the buss. She didn't know just where we lived so she got off at McKinsy's so she rested a while & walked on down here. I thought it was her soon as I saw her. She is kinda feeble, tho she should be, she's 60 years old.'

"Joyce said it was rumord that you boys were in South America. I'll be glad when we hear from you, tho I have learned to be patient or try to.'

"The war news sounds better everday, tho it seems like they have a long way to go. That is if the Germans don't give up & save their selves.'

"I imagin they are fixing to give the Japs another big blow pretty soon, around the Philipines.'

"H. D.s folks came up last night. Aunt Ella was with them. It's a pretty big job to get things fixed for V. Lee to go to Austin. She wants to go & she needs to for she is worrisome, tho she does help a lot with the work & I hate for her to be sent off. But I know she can learn a lot. They grade them according to how smart they are. She is anxious to learn to read & write. I hope she can.'

"Jack, Bryant bought the biggest car he could find, a Murcry 8. Gave his A ford & $975 for it. H.D. said he felt sorry for him, ho."

R. C. added a note to Mamma's letter,

"I am workeing over at the camp service club now. One of Louis Ladds brothers are workeing there in fact he is my boss.

"We got a letter from Herchel yesterday. He sent some pictures and some money for June's birthday.'

"Well I had better go to work. Be good and write soon."

Mamma wrote Jack again on August 19,

"I thought I would drop you a few lines to let you know we heard from your things. They were sent out the 12 of this mo. from Ft. Riley so they will be here soon.'

"I wrote up there to Ft. Riley & ask about them & told them to please see if they could find out anything about them. Told them to see your soperior officer. I don't know if that helped or not. I told Dady you might have written the Lt. that you left them with if you had heard from me. Any way I'm glad we had word from them.'

"Dad went to the Dr. today & he said he could go back to work Monday and am I glad to know he is that well. His boss is awfuly good. He will have him to be careful. He sent Dady a card when he was sick.'

"We got one of your $50 bonds today. Loyd sent one home last week.'

"We had another letter from Herchel yest. He said he could tell us where he is now & a lot of things he couldn't before. He's at Pearl Harber.'

"We sure will be glad when we hear from you & Curtis.'

"I think R.C. likes his job at the service club. I want him to stop & start to school when school starts. Mabe things will change & he won't have to go to the army.'

"I saw 2 soldier boys get caught yest. When we was at Rosas they came up to Mr. McCrackens to get some melons. They came by Aunt Rosas in a big hurry & I was feeling sorry for them and about time they got their melons a Major drew up & they droped their melons & he talked to them a while. So they took their melons back & came back by so I felt sorry for them & offered them 2 cantalopes. They said they had lost all taste for anything & wouldn't take them. So Jr. was at Elberts and there was another boy waiting for these two, so they got a little McDonald boy to go & get them 2 mellons. So they got some after all. I know they were out of place but I didn't blame them. Then there was 3 or 4 up at Marvins having music out at the garage. They ask Dad what time it was & it was 4 I think & Dad said they sure did get up & run back to the reservation."

Mamma's letters for September, October and November, 1944 are unavailable, presumably lost or destroyed. I think it miraculous that I found the ones she wrote during the summer of 1944 and later during the winter of 1944-45.

CHAPTER 14

INDIA AND MORE TRAINING, WITHOUT HORSES

LORD LOUIS MOUNTBATTEN, Supreme Allied Commander for the China, Burma, India theatre was formulating plans to open a line of transport and communications from Rangoon through Burma to the Chinese border to assist the Chinese in their war with the Japanese.

"While being pushed back in the South Pacific, the Japanese had continued a successful expansion into Burma toward India. This ended in June, 1944 on the plateau of Imphal. Three Japanese divisions were thrust back by British and Indian forces with great loss of men and equipment. The allied victories came slowly with great difficulty under monsoon conditions and in jungle-covered mountainous terrain.' (WWII, American Heritage)

American forces, under the command of General Joseph 'Vinegar Joe' Stilwell added to the Allied successes with the capture of Mitkyina in northern Burma on August 3, 1944. The capture of Myitkyina was important because it was the northern terminus of a railroad from Rangoon. It was also the head of river navigation on the Irrawaddy River. Just two miles from the city was the only all-weather airstrip in northern Burma. It was here during the battle for the city that a young Lieutenant Leo Tynan, a recent graduate of The Virginia Military Institute, flew in to join the Merrill's Marauders and earn the much coveted Combat Infantryman Badge. He would join others from the Marauders as they became a part of the newly formed Mars Brigade. This strategically important victory gave the Allies a staging post for the American Air Force to send supplies into China. Myitkyina would prove equally important to the Mars Task Force as a forward base for land operations. (Marsmen in Burma)

During this period of time, strategy was being formulated at the highest levels of leadership among the Allied governments, with some debate. The military needs of Europe, the political concerns in China and the strategic needs in Southeast Asia and the South Pacific were a problem impossible to solve and meet the goals of all theatres of operation. (WWII, American Heritage)

Lord Mountbatten journeyed to London to discuss strategy with Winston Churchill. They had agreed on a plan to capture Rangoon with an amphibious landing. This action, if successful, would save lives and accelerate the re-capture of Burma. The problem with the plan was that it required two additional army divisions. At the time, these divisions could only come from Northwest Europe. General Marshall turned down Churchill's request for U. S. divisions, sticking to the plan to beat the Germans before concentrating all resources on the defeat of Japan. (WWII, American Heritage)

The USS Butner sailed into the harbor and anchored at Bombay, India on August 26, 1944. Curtis wrote the following account.

"We stopped at Sydney, Australia to get supplies, fresh water and fuel, but no passes because we had some Negro troops aboard and they weren't allowed to go ashore, but just to stop that ship was worth a lot. We go out on the deck for a few hours. When we left Australia, we got on the Indian Ocean and headed for India. We had two destroyers and a cruiser for escort. We zig zagged all the rest of the way. When we crossed the equator again they started stopping up the vents up above us to turn the air into their compartments keeping the cool air away from us. Sure was hot.'

"We landed in Bombay, India. As we entered the harbor, it was so foggy that we couldn't see the dock until the ship stopped. We had spent 31 days aboard the ship. Got a bunch of mail from Joyce and my folks. That's the best thing that has happened so far."

The young Texans along with others who had been assigned to the 124th from all over the U. S. found excitement in the new and exotic surroundings. The dark-skinned people, the unusual dress, the numbers, the poverty, the filth, the strange customs were at first shocking, but were a welcome relief from the confinement of the troop ship.

With little time to be swindled by the Bombay peddlers, or contaminated by the prostitutes, the regiment was disembarked, loaded on the trains and carried east across India to Ramgarh, near the Burmese border. Jack wrote his first letter from overseas August 26,

"Well I finally got a chance to write. I don't know what I can say yet so I just won't say anything. Curtis & I are both o.k. We have been having some fun since I wrote last & it looks like it is just beginning. I think we will be able to tell you all where we are pretty soon. I can say that it is quiet an interesting place."

From Ramgarh, India, September 4, he writes,

"How are you all today? I'm fine. Curtis is o.k., too. I saw him about 2 hrs. ago. He was at a meeting we had. I see him every day.'

"Boy this is some country. I can't begin to tell you all the things I've seen. First thing I noticed was the very unusual dress. The men wear turbans & a sheet affair wrapped around their bodies. They usually have small thin legs, big bare feet & are as black as the ace of spades. They work for a little of nothing & eat very little. That is one class of them. You have a begger class & then the wealthy ones. Everybody is born into his class & stays there. It's not like we have where Joe Blow can be president if he's got guts & brains enough. They all seem to like the Americans very much & think we are all rich. That's because we give them things. Some of the Indian soldiers are the best in the world.'

"The money is different also. We have the rupee as a basis. It's 32 cents worth. The Anna 2 (cents) and the Pice about ½ (cent). I'll send you all some & let you see what we buy things with.'

"This is not a bad camp at all. I like it. The monsoons are about like our rainy springs.'

"I haven't heard from any of you all for about 4 days now. Your last letter was written on the 19th of Aug.'

"Oh yeah, about my junk. I turned it over to the Special Service Officer & he shipped it for me free. I guess we all got a low priority on shipping space & it will be a long time before it gets there. There's a foot locker, a barracks bag & a big box. You all use any of it you can.'

"Well don't worry about us. We are as safe as if we are home in bed. Besides, remember what I told you?...it would be quite an honor to have a son shot in this war. I'll have to deprive you of that honor if I can.'

"You all be good & write. I hope you are feeling good now Pop. You dang brats write too."

Apparently Jack's attitude about military service had progressed from impatience in late 1940 and early 1941 to complete a year of obligatory service to a fatalistic, sacrificial attitude by the time he was confronted with the reality of combat. Even though these cavalrymen were overseas and expecting to fight the Japanese, there were still delays and a tourism atmosphere.

On September 6 the USS Maryland was anchored at Port Purvis, Florida Island, Solomon Islands to take on fuel and ammunition for the task ahead. (NACP, N.G. 24)

Jack wrote September 10,

"I went to Ranchi today. It is pronounced like ranche. There is the filthiest hole I ever saw. R. A. Langley the trash hauler in Mineral Wells could be mayor of that town. It's a good sized place & you can buy lots of things if you can stand it long enough. The trip over there is worth a lot. It is about 28 miles of the most unusual country you could see. One stretch of about 5 miles is all going up hill from here & I'll bet there's over a hundred curves that almost double back to the same place. The hills are pretty high & there's a valley off to the left about half a mile deep that is matted with rice patties & you can see water falls on the hills across the valley. It is all green now that the monsoons are in season. The Indians are always working on the road. It is a hard surface, but is only a one way & you have to slow up to pass a car.'

"The Indians dig dirt out of the side of the hills above the road & carry it in baskets, one on each end of about a five ft. stick, & build up the road on the outside of the curves. They all stop & watch the trucks pass. Women work right along with the men.'

"One thing funny that I've seen here that you won't see at home, are the hawks. They circle all above our kitchen while we eat & come down like dive bombers & grab things out of the boys mess kits. They won't even fool around when we aren't eating.'

"I got some stuff for you all in town today. I'll send it as soon as I can. We can send one 10 lb. package per mo.'

"Well the news sounds pretty good, but I'm not getting optimistic, because it has been good now for about 2 yrs. I'll send you all some Indian money in this letter if I don't forget it.'

"Be good & write soon & often. Mom you can do as you like about alternating our letters. We are close enough together."

The 124th Cavalry Regiment had become a part of the 5332d Brigade (Prov). The regiment was in the process of training for conversion from a mounted unit to a special force much like the infantry. On August 26, Training Memorandum Number 2 was issued with the following summarized directions.

"The objective of the training program from August 28, 1944 to September 30, 1944 was to bring personnel to a high state of basic training to function within their units from the squad up to the battalion level. The men were to receive special emphasis on care and use of individual weapons, map reading, individual specialty techniques and physical conditioning.'

"Unit training was to include combat firing, reconnaissance and security, decision, decision- making and command functions, march discipline and communications. The training week was to be a minimum of 36 hours. Regimental and separate organizations like the artillery units were left to do additional training at their discretion. They would suspend training on Sunday."

(Series Mars Task Force, World War II, Operations Report, Entry 427, Record of the Adjutant General's Office, R. G. 407, National Archives at College Park, Md.)

Proper training was extremely important for the success of the mission and the cost of adequate training can be illustrated by presenting the requisition of ammunition from the 475th Infantry for the month of September from the archives.

81 MM Light	1,640 RDS
81 MM HE	300 RDS
60 MM HE	4,000 RDS
60 MM ILLUMINATING	400 RDS
60 MM SMOKE	200 RDS
GRENADES, AT, RIFLE	1,500 EA
ROCKETS, AT	500 EA
CAL. .45 BALL	108,000 RDS
CAL. .30 MG	65,000 RDS
CAL. .30 8 RD CLIP	150,000 RDS

CAL. 30 CARBINE	20,000 RDS
CAL. 30 TRACER	5,000 RDS
SHAPE CHARGES	25 EA
TNT	500 LBS
PRIMER CORD	600 FT
FUSE	1,000 FT
BLASTING CAPS	2,000 EA

This requisition was for one infantry regiment of just over 3,000 men for one month. Extend that to the whole war effort and add some for waste and we see that war is ravenous as well as hell. Necessary administrative decisions were being made as the troops moved to their training facility.

A memorandum confirming the decisions reached at CHABUA Conference was written and distributed on September 5, 1944. The following excerpts have been lifted from that document to shed some light on the formation of the Mars Task Force, especially the changes wrought for the 124th Cavalry Regiment.

"The 1st Infantry Regiment, Separate (Chinese), will operate as a part of the 5332nd Brigade (Prov). It will be moved from RAMGARH TO LEDO on October 15, 1944 or as soon thereafter as practicable.'

"The 124th Regiment will be expanded into a three-squadron regiment. To accomplish this, cadres for three squadrons will be formed from the personnel now comprising the 124th Cavalry Regiment. If required in order to reach full strength, use of personnel from the initial group of monthly replacements (55 officers, 600 enlisted men) arriving in late September is authorized.'

"Without exceeding the strength, grades and ratings which will shortly be authorized for the 124th Cavalry Regiment, a regiment similar to the 475th Infantry Regiment.'

"On arrival in the theater in late September, the initial group of monthly replacements for 5332nd Brigade (Prov) will be shipped to RAMGARH for 124TH Cavalry.'

"As a basis for preparing a new T/O for the 124th Cavalry Regiment, grades and strength will be as prescribed by T/O 2-11, 1 Apr 42 with changes (81 officers, 1571 enlisted men), plus one Cavalry Rifle Squadron, T/O 2-15, 1 Apr 42 (20 Officers, 495 enlisted men), 1 Machine Gun Platoon (1 officer, 40 enlisted men), and a mortar section (17 enlisted men) per column 7 less 2 basics and column 100, T/O 2-19, 1 Apr 42. Total grades authorized, 102 officers, 2123 enlisted men, total 2225.'

"Although the War Department has approved the addition of a third Squadron to the 124th Cavalry Regiment, actual War Department authorization of grades and ratings for a third squadron has not yet been received. Until received, grades and ratings in excess of a two-squadron Cavalry Regiment per T/O 2-11, 1 Apr 42, with changes are not authorized. All concerned will be informed by radio when the War Department authority to use the additional grades and strength is received. Pending receipt this authority, reorganization will proceed but promotions to fill new T/O vacancies will be held in abeyance.'

"By command of General STILWELL:

EDWIN O. SHAW

Lt. Col, A.G.D.,

Actg. Adj. Gen."

(NACP, R.G. 407)

Jack was later caught in this squeeze. He would command a troop, without the usual captain ranking.

On September 28, 1944, GENERAL ORDERS NO. 4 was issued from Headquarters of the 124th Cavalry, (Special) for the reorganization and redesignation of the regiment. It stated:

"The reorganization and redesignation of the 124th Cavalry as the 124th Cavalry, Special, effective as of 25 September 1944, pursuant to General Order Number 124, Headquarters United States Army Forces, China Burma India, dated 25 September 1944, is hereby announced.'

"Concurrent with the above, activation of the 3d Squadron, 124th Cavalry, Special, effective 25 September 1944, with an authorized strength of 20 officers and 495 enlisted men, was directed and is hereby announced.'

"Pursuant to the above quoted authority, activation of the following units of the 3d Squadron, 124th Cavalry, Special, is announced:

Headquaters 3d Squadron, 124th Cavalry, Special

Troop I

Troop K

Troop L

By order of Colonel Matteson:

DELBERT M. TANNER,

Major, Cavalry,

Adjutant."

(NACP, R.G. 407)

At home we were trying to absorb the changes in the family structure and to deal with the rapidly developing status change for the Knight boys in the war. We had started back to school early in September. Our summer activities overlapped the activities at school. It was about this time at home of the 'peach raid'.

Our adventures around the Brick House continued late into the summer. By now it was only Roy and me involved in these escapades. We roamed the woods at ever increasing distances from the house. On a particularly extended excursion we came upon a peach orchard. I don't recall the variety of peach, but even at this late date there were still a few peaches clinging. We must have had better than usual summer rains to encourage this late growth.

Quickly we rationalized that the farmer had gotten all the peaches he wanted and didn't need these few remaining. With no attempt at cover or concealment, we helped ourselves to a minor feast of these ill-gotten fruits. Without guilt or anxiety, we stuffed a few of the firmer ones in our pockets to eat later and share with our parents, R. C. and June. We proudly and innocently produced our booty and placed the dozen or so peaches on the dining table.

Mamma and Daddy, the ever-vigilant guardians of our conscience, morals and souls began the inquisition.

"Where did you get these peaches?"

"Over there about a mile." Roy said, pointing to the northwest.

Daddy knew immediately whose orchard it was.

"Did you ask Mrs. McDonald if you could have them?" Mamma inserted.

"Oh, my Lord!" I thought to myself. She was principal of my school. I just knew my butt was mud. We had been in school only a few days. I didn't know her that well, but she was the elegant and businesslike lady who was Principal. I knew what was coming because I had seen Mamma and Daddy double-team my older brothers on prior misdemeanors.

"Land's sake, Junior…Billy Rowe. You know better than to take things that don't belong to you. Why did you do this?"

My mouth became cotton and no sound could be produced, much less a response. Roy managed an almost inaudible,

"We didn't think nobody would care. They wuz only a few left and they wuz fallin' off and layin' on the ground."

"That don't matter. They wasn't yours." Mamma said.

Daddy spoke up,

"O.K., this is what I want you two to do. Take those peaches and feed 'em to the hogs. Tomorrow you will go tell Mrs. McDonald what you did and apologize."

Oh, Lord! How could it get any worse. I wished they would have whipped our butts and forgotten it. It was bad enough to throw away these perfectly good peaches, but to have to face my principal with this was a mind-boggling nightmare. In my mind I built up a scenario that would rival thumb-screws and water torture. It would be at least a thrashing with that board that hung by the black board.

After a restless night and an unwanted breakfast and too quick a bus ride to school, we entered the school building and walked very slowly to the room. As we entered, Mrs. McDonald was at her desk in front of her room. She also taught seventh grade. It had to be now or never and I knew that it could never be never.

With my patented red face we approached the desk and the dreadful moment with infinite trepidation. Mrs. McDonald looked up. Somehow, I managed to speak first even though Roy was the oldest and bravest. I opened my mouth, the blood drained from my face and my freckle spotted pale face opened with the feeblest of voices.

"Miss McDonald, Junior and me took some of your peaches yesterdy. I'm sorry that we didn't ask you first." I had done it and I felt better already. Nothing she could do would be any worse.

She knew immediately that my parents were the force behind this act of contrition and could also feel the depth of my guilt and recognize the sincerity of my apology.

"Billy, I'm glad you came and told me. That is the right thing to do. I don't think you will do anything like this again, will you?"

"No ma'm."

"What about you, Junior?"

"No ma'm, I won't. I promise."

"Now, you all go on to your rooms and we'll forget all about this. Alright?"

A duo of "Yes ma'ms" and the whole affair was over.

I smiled at that sweet and understanding face and did as she said. At that moment Mrs. Grace McDonald became an idol, a model to me. I worked harder that year than ever before. More than thirty years later, after I had earned my PhD and had become a school superintendent, she attended one of our class reunions. I reminded her of this incident. She remembered and chuckled. She told me that at that moment she knew that the Knight boys would be good men because of our parents.

Sight-seeing

India

With Chinese soldier

Mineral Wells boys in Burma

CHAPTER 15

WAITING AT RAMGARH

DURING THE LATE summer and into the fall of 1944, the Allied forces in Europe were inexorably closing the vice on Hitler's Germany. The Russians were building momentum for their rapid drive to run the Germans from their homeland and for the inevitable conquest of Germany itself.

The 101st Airborne Division was gaining respect for its courageous actions against the Nazi army. General George Patton was gaining ever more notoriety with his unrelenting pursuit of the Germans with his Third Army. The two units would gain everlasting fame for their combined heroics at Bastogne, which turned back the last real threat to the Allied advance through Germany. (WWII, American Heritage)

The USS Maryland steamed from Pearl Harbor to Peleliu to assist the Marines in that horrible battle, all the while fighting off attacks by Japanese warplanes.

The Japanese were holed up in caves and the Marines had to resort to napalm and flamethrowers to root them out. The Maryland then moved on to Leyte Gulf. (NACP, R.G. 24)

Jack wrote from Ramgarh, India on September 14, 1944,

"What are you going to do Jew? You should go to school & probably are. Don't start fooling around wanting in the army or navy now because you can't tell when this thing will come to a close. Wait till they ask for you. You know you need to go to school.'

"Have you all been doing any fishing lately? I'd sure like to be there & get ready for those ducks when they start coming in. That lake should be a hunter & fisherman's paradise when it is over." It was.

All of us were concerned about R. C. He had little interest in school. This problem was compounded by our move to the Millsap School District. It had been difficult for him to see himself as a Millsapper. His return to Garner was not accompanied by a changed attitude toward academic achievement. He spent most of his time thinking of excuses to miss class or simply walking out the door with the principal asking him to come back. He finally gave up the ruse and quit to go to work at a dairy near Weatherford and later at Camp Wolters while he awaited a call for induction. Jack was trying to use his considerable influence to keep him in school.

Again he wrote from Ramgarh on September 17,

"How are you doing now Dad? Are you still working? Boy it's really hard to get any decent news over here. We listen to these Indian radio broadcasts from Bombay & Calcutta & all around here & we sure can't tell what to believe. Then we get Jap newscasts and they try to make us think we are about to lose the war. They really play up this old bull about Asia for the Asiatics & cuss the English & Americans for butting into their business. They try to stir the Indians up against us but I'm sure that if they could see this country like we do they wouldn't waste any more time. The Indians don't want the Japs any more than we would at home. They have the biggest army of volunteers in the world, about 2,000,000. I guess the Japs have to talk about something tho."

As training continued with the various contingents of the Mars Task Force, they drew on the experience of the Merrill's Marauders and the British who had been fighting the Japanese in Burma since the outbreak of the war. The following list of tactics had been gleaned from these jungle fighters. The following statements are excerpts from a document contained in the history of the Mars Task Force located in Archives II in Maryland.

"Every man should have some guide or pamphlet to get information from natives – an interpreter for each platoon should be provided if the natives actually go with the Combat Team on independent operation."

"Marching at night in the jungle is very difficult for men over good trails – over fair or poor trails it is almost useless and is practically impossible for animals."

"If you are in or near enemy territory don't walk around at night even to relieve yourself. There is always a trigger happy guy who shoots first even though you are miles from the enemy."

"We drove a mule in front of us over a trail we thought might be booby trapped. It was, and the mule was killed."

"Insist that men don't dig foxholes near trees – either dig away from them or cut them down – tree burst are responsible for many unnecessary casualties."

"Some of the most valuable men in our outfit were the Nissi Japanese interpreters not with battalion and regimental headquarters but with a platoon in contact on the perimeter. The Japs talk loudly sometimes before they attack – On several occasions the Japanese interpreters told us exactly what the Japs were shouting and enabled us to get set for an attack from a certain direction. Once an interpreter caused the Japs to attack into a trap by shouting orders to them."

"It is easy to teach a man to look for different type shoe prints on a trail. On two occasions a suspicious looking shoe print caused us to surprise the Japs whereas if we hadn't noticed it they probably would have surprised us."

"The Jap is no suicide soldier – He will run and retreat if surrounded and once you get him on the run he is not nearly as effective."

"One of our patrols took the same route that a previous patrol had used and were ambushed. The Japs are cagey – they will follow a patrol sometimes for long distances then set an ambush in case it comes out again." (NACP, R.G. 407)

This is only a partial list with only a few items that were covered in training. It demonstrates, however, the minute detail that was covered in order to gain an advantage over the enemy.

Jack, a first lieutenant, remained in the headquarters unit and Curtis was still assigned to Troop F as mess sergeant. The regimental units were no longer scattered hundreds of miles along the Rio Grande River. The men of the 124th were now located together at a training base near Ramgarh. 'The Boys' had opportunities to meet and visit daily.

Curtis decided to go to the PX to break the monotony and see if they had anything worth buying. He dropped by Jack's tent.

"Hey, lieutenant, I'm going to the PX. You wanta go?"

"Yeah, just a minute. I've got to get my jacket buttoned so I can look like an officer." They set out through the slush created by monsoon rains.

"Well, John, what do you think of this place?" Curtis inquired as they slogged along.

"I never expected to wind up in this neck of the woods. I thought for sure that we would be sent to Europe. Cavalry could be more effective there."

The horse cavalry mind-set was hard to break, notwithstanding its lack of use in the war. Up to that point there had been no direct indication that the unit would cease to exist as a cavalry fighting force.

"Do you have any idea what we're in for?" Curtis asked.

"Not exactly. You know that we'll wind up fighting the Japs. The sooner, the better."

"Yeah, I guess so." Curtis agreed. With thoughts of Joyce and Glenda, he wasn't as anxious to mix it up with the Japanese as the impatient Jack.

Jack went on, "I've been chomping at the bits for three years, waiting to get a chance to kill some of those slant-eyed bastards. Now I'm sitting on my butt at headquarters. I still might not get my chance." They walked on in silence.

"What are you looking for?" asked Jack as they entered the PX.

"Nothing in particular. Just thought I'd check to see if they had any candy or other things to eat. I don't need anything."

They picked up a few bars of candy, paid for it and left. There were no cold drinks to buy.

"Have you heard from the folks lately?" asked Curtis.

"Yes. I got two letters today. It only took 'em a week to get here. The kids are getting ready to start back to school. I don't think Smutter is going to stick in school. He may not make it through his senior year. He couldn't get used to being a Millsaper and he won't do much at Garner. He'll probly join the army before the year is up."

"You're probly right. Darky, Gopher and June are doing o.k. at Millsap. They never had to play basketball against those smart-aleck Millsap jerks."

Millsap was located about six miles south of Garner. It was a larger village and had more students in school and a more comprehensive academic program. They had a gymnasium and could play their basketball games inside. They even had tennis courts. Most of the time the Millsap teams prevailed over Garner, resulting in a bitter rivalry bordering on hatred. The Garner boys tried to get back at them after graduating by trying to steal their girl friends, with modest success.

As they approached Jack's tent, Curtis asked, "How long do you think it'll be before we get to fighting?"

"Probly a few weeks. We've got to do some training and get better organized. I have no idea what the brass is planning. You know we have a Limey Lord in command of the CBI. We won't have any problem finding Japs, though. They're all over this part of the world. I guess we'll find out before long."

Jack and Curtis continued their stroll back to their unit in 'Tent City'. Even though near the end of the monsoon season it continued to rain with resulting miserable living conditions, mud outside, mud inside. It was very difficult to adequately ditch water around so many tents and the water ran right through them.

They decided that the first time they could get off, that they would take one of the trucks provided by the army on an excursion to Ranchi for sight seeing and souvenirs. Jack had been there and he wanted Curtis to see the sights.

They would also have an opportunity to make a trip to the Majaraja's palace, a disappointment. It looked too much like an American resort hotel.

"I've heard rumors that our horses are being shipped from Riley." Curtis said.

"Yeah, I've heard that too. I've heard that horses are coming from Australia and India. I don't know. We'll just have to go along, follow orders, do our training and let the chips fall. I really don't care as long as we don't waste too much time. I've waited long enough. More training after four years of it won't help me kill any more Japs."

"I guess so. We'll know something before long. I've heard that Lord Mountbatten, even though he's a Limey, is pretty smart and real tough. He's ready too, more than likely," added Curtis.

The fantasies of regaining their mounted status soon evaporated as the 124th began making the transition to a specialized jungle-fighting unit of foot soldiers. They were required to turn in their mounted equipment. Riding breeches were replaced by infantry fatigues, shoes & leggings were issued; riding boots and spurs were turned in. There were no more rumors about the arrival of horses and no more illusions about the fate of horse cavalry. The appearance of mules puzzled the troopers awhile. Their temporary anxiety over these animals serving as mounts soon dissipated with the realization that they would be used as pack animals. Japanese to the east. Mountains. Pack animals. Training to fight in mountainous jungles. Training to cross rivers and streams. The vision was clearing. Combat was just around the corner. (Marsmen in Burma)

Soldiers respond to the pressures of war in diverse ways. Some drink alcohol excessively. One F Trooper whom we will call Cowboy, had an active acquaintance with the spirits, but India provided another taste option, as well as an excuse. "Bamboo Juice", the drink of choice for the Troop F practitioners of over-indulgence was a rice saki concoction that was capable of quickly knocking one on one's butt.

Cowboy, a free spirit with little positive regard for military courtesy, regulations or discipline of any kind, was one of the first to discover the sweet numbing effect of this beverage.

Curtis was mess sergeant at the time and discovered Cowboy in the throes of his recently discovered anesthetic. His usual disregard for sensitivity had been enhanced by a jug of 'Bamboo Juice'. He was proclaiming to all who could hear in a most profane manner that he considered his present state of existence to be quite unsatisfactory. In broad daylight, in the middle of the day, this means of personal expression is frowned upon in military surroundings.

Curtis, being the good and responsible sergeant, found his choice of words and manner to be offensive as well as a violation of military decorum. He attempted by gentle persuasion to calm him and convince him to adjourn to his tent to get himself under control. In his drunken state he took the diplomacy of Curtis to be an aggressive attack upon his freedom of inebriation. His attempts to express his displeasure at Curtis's affront were ill advised. Curtis quickly discerned that reason was beyond Cowboy's present mental capacity. If not reason, then what? He popped him with a short right cross to the jaw, kept him from falling by grabbing him around the waist and carry/dragged him behind the latrine and pushed him down. Unfortunately, he staggered into a water-filled ditch and landed face down in this mixture of muck laced with human excrement.

Curtis had to go in there and extract him to prevent him from drowning and to avoid the necessity of procuring a very scarce combat replacement. This in turn required him to drag the private back to the shower and a good bath for both. By the time all this was completed, Cowboy was under a semblance of control and Curtis was on his way to supervise the preparation of the evening meal. This was just another day in the life of young men preparing to kill and to be killed.

Cowboy's reputation for contempt of military life went beyond this incident and followed him into Burma. As training continued at Ramgarh, the F Troop commander, Lieutenant Jack 'Snake' Warner was conducting a lecture for his men when one of his pronouncements elicited a contemptuous laugh from our friend, Cowboy. The lieutenant just stared at the private for a few seconds and went on with his lecture.

The training session was completed and the men started back to their tents. Lt. Warner walked up beside Curtis, who was attending the training sessions, even though he was mess sergeant.

"Knight, I want you to beat the shit out of that guy. He really chaps my ass. He has been a pain in the ass since Ringgold."

Curtis had had his round with Cowboy and answered, "Why should I fight him. I don't have an argument with him. I can't do that."

"Well, I can't either. What would it look like for an officer to hit one of his enlisted men?"

Nothing else was said, but his animosity toward Cowboy continued to simmer. It would surface later with results that the 'Snake' would find unpleasant and ultimately regret. 'Snake' came from his boxing ability; his quick fists were said to strike like a snake.

Jack wrote from Burma on October 3. Apparently he was part of the advance party for their move from Ramgarh.

"Well we just heard today that the Browns won the Amer. League pennant. I can hardly believe it. Well they and the Cards should make around a rupee each off the series.' [The war had a profound effect upon the landscape of major league baseball. The usually hapless Browns found the effect equalizing with the loss of so many stars from the stronger teams. The two St. Louis teams may not have made much money, but they had a decent series, the Cardinals winning four games to two.]

"Oh yeah…did you save any newspapers for me about the European invasion. Be sure & do that & save some when the Germans quit & anything you see about us…I mean the outfit."

There would be more written about Troop F and 'The Boys' than any of us could ever have expected or wanted.

He wrote again October 7,

"Sure wish you could be here in peace time. This is a heck of a good country. The mts. are beautiful but I guess they are pretty rough to do any traveling in. It should be fun for Pot Guts & me tho. The swimming holes are thick but I guess winter will cut that out pretty soon."

Burma Campaign

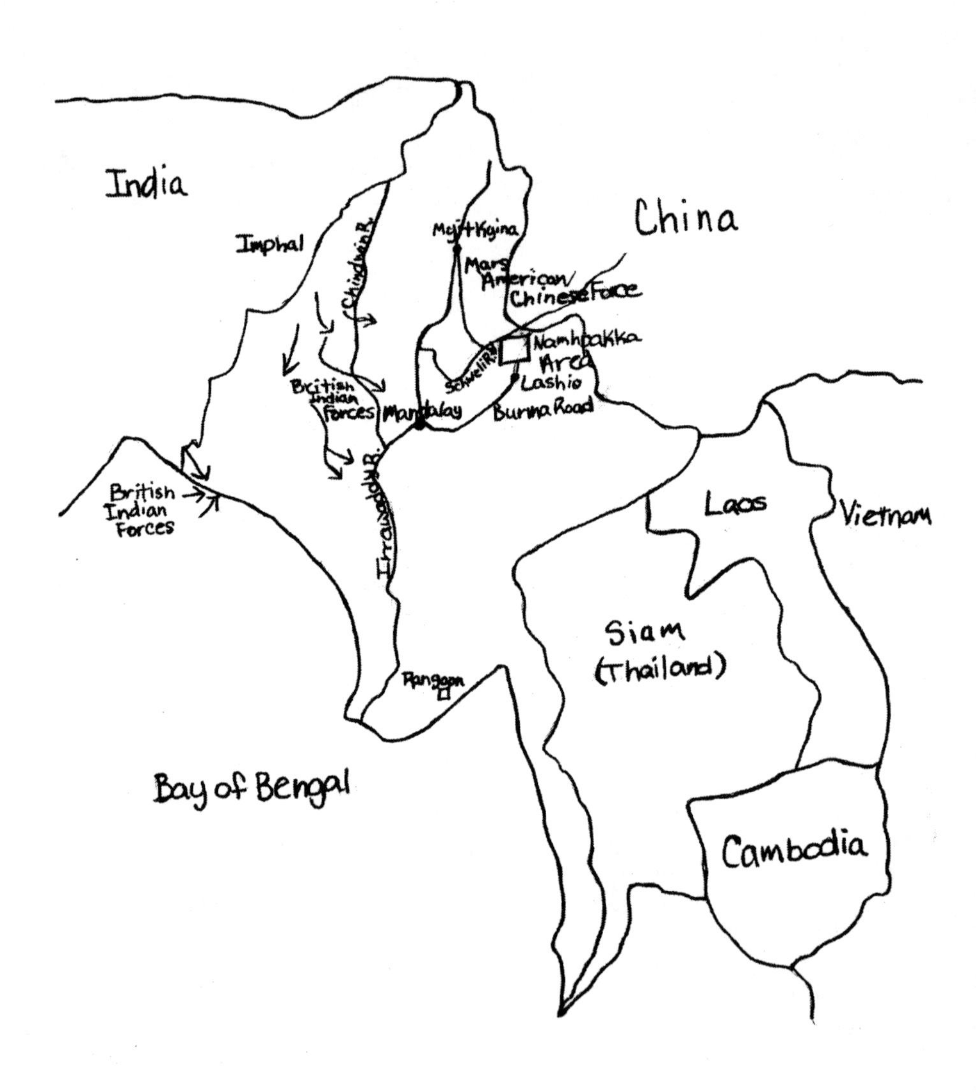

CHAPTER 16

CAMP LANDIS, BURMA

THE 124TH LEFT RAMGARH on October 15, 1944 and traveled to Burma. The two-week trip included various means of transportation: standard and narrow-gauge railroad, side-wheel river steamers up the Brahmaputra River in the vicinity of Ledo, by air on C-47 transports to Myitkyina and then by bus to Camp Landis. (Marsmen in Burma)

General Douglas McArthur waded ashore at Leyte on October 20, 1944 to fulfill his promise made when he was forced to leave Corregidor on March 11, 1942. (WWII, American Heritage)

The 475th Infantry had built the camp, which had been named for Private Robert W. Landis, the first man of Merrill's Marauders to be killed by the enemy. The 475th had been formed from remnants of the Merrill's Marauders and other infantry units and other assorted recruits. Four hundred mules arrived November 4 via the Ledo Road, a dirt and gravel highway being constructed by the Allies from Ledo in Northern Burma southeast to the Burma Road, passing near Myitkynia. Other mules were brought in with specially trained pack units. These two regiments along with other support units were placed together to form the 5332nd Brigade, also known as the Mars Task Force.

Camp Landis would serve as a staging area for the newly created brigade. The 124th would have a name change to the 124th Cavalry Regiment Special, dropping the designation of 'horse'. They would be further trained as infantry.

Their mission and the tactics for its execution were beginning to become obvious. Their location was near the Ledo Road that led to the Burma Road, which meant that the Japanese were close by to the east and south. Hundreds of mules were arriving in Camp Landis. There were few motorized vehicles. Many of those were the result of some enterprising refugee mechanics from the old motor pools. They managed to piece together some operable vehicles from wrecked Japanese and British trucks that had been abandoned in the area during recent fighting.

Kachin Rangers along with some very courageous American scouts reconnoitered the mountainous jungle territory to the south and to the west of the Ledo road. Their report included recommendations on the distance that could be covered each day, features of the terrain, feed available for the mules, water supply, firewood and air drop locations. They would march 6,000 troops and hundreds of mules through this region much to the chagrin of the Japanese. Tokyo Rose would keep the troops informed of their movements, but the Japanese intelligence vastly miscalculated the numbers of Americans on the trail. (Marsmen in Burma)

Jack was trying to let us know what was ahead for them. They were facing this formidable obstacle that an army could cross only on foot with pack animals.

He wrote October 9,

"I've got a Jap dog tag. I'll send it home to you as soon as I can find out if I can or not. It's something like ours but it's got some hen scratching on it that won't make sense to you. No, I didn't shoot the little _____ that wore it."

Impatience was again developing. He had waited years for a chance to go overseas to fight. He was overseas, but was having to wait again. He wrote again on October 15,

"Well times a wastin'. I'm ready to go do some fightin' now. Pop you want to study up on your Burma geography & get you a map so you can keep a situation map. It might be fun."

Jack couldn't tell us what their military objective was but Daddy knew about the Burma Road and knew about its strategic value.

Curtis wrote from Ramgarh on October 11,

"We are doing a lot of hard training here. They made me Platoon Sgt. Bill McQuerry took over the kitchen."

Since Jack had gone ahead to reconnoiter their camp site, he wrote again from Burma on October 23,

"I think Buzz is pretty fed up with his present situation. He thinks he'll never get into the war, but he can't tell. He may be in it before he knows what's going on.'

"I'm sure glad to hear that you kids are going strong in school this year. You might as well go on & study hard & learn all you can, because you are all going to college if I have to ear you down & make you go.'

"How tall are you anyway, Jew? Dang, you should be able to take on a whole basketball team by yourself. Who all is on your team this year. I wish you could win the county tournament this year. We never could, but we never had enough Knights playing at one time.'

"How are you baby June? Or do you think you are too big to be called baby. I don't think so. You either Gopher. You are all sweet in these pictures & I love the hell out of you.'

"What have you done about your job Dad? I know how tiresome it must be riding all that way.'

You all start looking around for a place around the $5,000 class & when I get home we'll buy it. I'm going to need a fishing & hunting headquarters."

The USS Maryland had reached the Leyte Gulf by October 19 and at 1301 had begun firing its secondary to port on a target in the vicinity of Palo. Four days later the large naval Battle at Leyte Gulf began. On October 25, the Maryland was lying in the waters off Leyte, guarding the straights. The Jap fleet had been reported headed for the San Bernadine Straights. She maneuvered to pick up the pilot of an F6F that had been shot down by a Jap fighter. The deck log of the Maryland contained the following entry, dated October 25, 1812.5 hours.

"Observed enemy plane diving on this ship. 1813 crashed into Turret One, bomb fell between turrets one and two, piercing the deck and exploding on second deck. Plane on fire, port side of forecastle. Ceased firing AA battery. 1816 Fire extinguished. 1827 Changed course to 310o T. 1852 Changed speed to 2/3, 10 knots. Set material Station Yoke aft of frame 50. Damage done to the ship not as yet ascertained."

(NACP, R.G. 24)

The Maryland was hit by a Kamikaze. The introduction of this suicide weapon was a dreaded new wrinkle to which the navy had to adjust.

On November 2, Jack wrote,

"You will mow me down for not writing, but I've been doing other things that take up a lot of my time & now that I'm back in F Tr. & having to censor mail again I'm swamped. I'm glad to be back in old F Tr. It's just like coming back after being gone from home...Well not quite that good, but you know what I mean being as how Curtis is here now."

Soon after arriving at Camp Landis, Curtis was promoted from staff sergeant, a platoon leader, to first sergeant, skipping the rank of sergeant first class. This made him the highest- ranking non-commissioned officer in Troop F.

Continuing his letter, Jack wrote,

"He's playing 1st Sgt. now. Sgt. Whatley got messed up and they broke him. Curtis will do the job o.k. & if he makes his rating it will mean a lot to him & Joyce & their bank account as well. Its one of the best jobs in the army. I'm executive officer of the Tr. I had a little stretch of duty as S-2 of the Squadron, but I never told you about it. I like the job but I like to have an outfit I can train & fight with. I think I'm gonna be the meanest man the C.B.I. has ever seen. I may turn out to be the scaredest & fastest on foot too. I'll just have to wait & see.'

"I heard from Toughy yesterday. He is still going strong. He's already seen more fighting that any man could ever see on the ground & live. That Navy is the place to be if you want to see action. From the looks of things now ours is the best by far.'

"Mom I ain't gonna tell you anything or put any Xs in no corners. Now you just settle down little gal. If I get shot you'll hear about it 'cause I'm wearin' my dog tags all the time.'

"It's not any 120° here. It's nice weather & dang cool at night, two blankets worth.'

"I was sure glad to hear from you Dad. You should make a cleaning on your pecan deal. Be careful about climbing tho. Don't think of it. You know you aren't any kid any more & remember what happened to Grand Pa.'

"June you are certainly improving in your writing. I enjoy you kids writing too much. You are going to have a pretty hand writeing June if you take your time & practice. Too bad about making 0 honey. Get Mom to get you a piano & I'll send you the money. I'll get rich now that I'm up here. I haven't drawn any pay for over 2 mos., & haven't had a chance to spend but 8 rupees, for envelopes & candy during the past six weeks."

Jack could be brutal in his messages, but he felt he had to prepare Mamma for the worst news. He also dealt very directly with Mamma's schemes and anxiety.

CHAPTER 17

JACK AND CURTIS TAKE COMMAND

DURING LATE FALL and early winter, 1944 the allied forces were at the door of the Third Reich. They captured Aachen, the first German city to fall. Operation Market Garden sputtered in an attempt to establish a port to land supplies near the front lines. The Allied advance on Germany halted and allowed Hitler to make his last desperate attempt to regain offensive momentum. On December 16, 1944 he sent his armies, to the complete surprise of the American and British commanders, into Belgium for what came to be known as The Battle of the Bulge. By the end of January the Allies had driven the Germans back to their previous positions with great loss of fighting men and equipment on both sides. (WWII, American Heritage)

At the same time the battle for the liberation of the Philippines raged. The Japanese were systematically and relentlessly being driven from their captured territories. Their removal from Burma was gaining momentum. (WWII, A.H.)

The Japanese controlled the northern section of the Burma Road. The British with some help from American troops needed to clear the Japanese out of the vicinity of this passage to secure an overland route from the ports of South Burma to China. The road ran over thirteen peaks exceeding 2,000 meters. The highest of these peaks is Tienatze Miao with an elevation of 9,200 feet. At some places along the road, a cobbled road five or eight feet wide crossed or paralleled the present Burma Road. Portions of this old Chinese road were estimated to be 2,000 years old. (NACP, R.G. 407)

Quoting from army records relative to its construction,

"Building on the road which started about 1920 was without engineering as we know it. It was done by coolies on the original concept that a road was anything a truck could drive over in dry weather. By May, 1942 when the Chinese made their defensive stand at the Salween River and blew up the bridge the highway was 3 to 9 meters wide. Paving was 3 to 5 meters wide and consisted of a layer of 6-inch stones, placed by hand, covered with 2 inches of smaller stones as a wearing surface. In addition, short curves and bad curves were pitched (like Belgian block), with this also placed by hand.'

"When Americans started cooperating with the Chinese National Government's Yunnan-Burma Highway Engineering Administration in early 1943, it was essentially a one-track, alweather road. Maximum grade was 21 percent with numerous short grades of 5 to 50 feet exceeding this, and, approaches to all bridges, located in sumps, exceeded this grade. Minimum curve was 25 feet on the center line of the road. In mountainous and rocky sections the road was as narrow as 8 feet." (NACP, R.G. 407)

The Mars Task Force had been assembled and trained together at Camp Landis. It was time for them to think about moving south. Jack wrote November 8, 1944,

"Curtis & I both had letters from you all & Herchel today. He is doing fine. Talking about the world series & still hates the Japs so I guess he is o.k. I haven't heard from Buzz now for about 2 weeks.'

"Well, I'm back in F Troop now. I like it pretty good. We have a good 1st Sgt., Pot Guts. He's really a tough top kick. He doesn't know how to be mean to anybody. You know that. I'm sure glad he got it. Joyce will get $205 per mo. now he says. Good stuff I'd say."

Herchel had had his closest brush with death as the Kamikaze hit his ship at Leyte Gulf. American troops had been surprised by the sacrificial nature of Japanese military tactics, but the American sailors, with their intrinsic high regard for life, were dumbfounded by the suicide planes being thrown into the Japanese defense.

Jack was upbeat now that he was back in Troop F. He wrote November 20,

"I've just got time to write a little but I'll slow down for a while & write. I'm working on a new job now & it is really keeping me busy. I like it tho.'

"Got your letter today telling about June's grass skirt & the haloween party. I'll bet she was a dillie. Spank her little but for me Mom.'

"Gopherstin why don't you get you some traps & make you some money out of those fox. Boy I'd like to be there to help you. I will pretty soon maybe.'

"I'm glad you all are doing o.k. in the pecans. That is a good job for you all. Don't let anybody fall out of a tree.'

"Well I gotta go now. Write soon & often.
Love John"

Having enough time to do his job and keep up his habit of writing every few days was becoming critical for Jack. He wrote again November 25,

"Well I got two minutes time & its 10 p.m. now to write. Since I've got my new job I don't have time to turn around anymore. I really have to keep on the ball.'

"Say we aren't where we can buy you all anything for Xmas so I'm just sending you all a money order. It will be $5.00 for each of you & Joyce & Glenda. Don't spend it all in one place. Just thinking of the future is the reason I'm not sending more. Oh yeah June baby when I get paid & collect some money for debts I'll send you some to get your piano. I should get it soon.'

"I know you are enjoying Joyce & Glenda being there. Curtis was just wondering how long you kids would let her stay without running her off. Smutt you'd better not drop that brat [Glenda], Murphy will send you a boobie trap.'

"Joyce I got your letter & was glad to get it. Don't let those kids get your goat. They aren't really as mean as they might seem."

Jack sent the money and June got her piano. She had lessons in classical piano for years. By the time she was thirteen, she had become an accomplished pianist, concentrating on religious music. During the summer of 1948 she played for a revival at Soda Springs Baptist Church. They passed the offering plates for her and she made over thirty dollars. I thought she was getting rich. She attended the Stamps Quartet School of Music in Dallas and has spent her lifetime playing in churches and for various singing groups.

On December 10, Jack wrote confirming his new position,

"Curtis & I are doing o.k. We have a good troop now. I really enjoy working with him & all the other boys from Mineral Wells. A lot of people are trying to transfer back into the troop now that I have it. I guess that's bragging but it's the truth."

Incredibly, four years almost to the day after 'The Boys' joined Troop F as buck privates, they were now the top officer and top non-commissioned officer of the troop. This was not singularly significant on a grand scale, but indicated how the United States was able to carry on so well during the war by flawlessly replacing leadership.

Our democratic approach to life encouraged emerging leadership from the ranks. Totalitarian, militaristic societies simply were not equipped socially and psychologically to accept leadership from anyone except those who had been assigned by tradition to the officer corps.

The Mars Task Force had remnants of Merrill's Marauders assigned to it.

Excerpts from Marsmen in Burma by John Randolph gives a brief glimpse of this courageous and efficient fighting force.

"It was decided at the Quebec Conference that the United States should have at least a small unit of active combat troops in this theater, and there were excellent reasons to have them here.'

"A call was sent out for volunteers, 'of a high state of physical ruggedness and stamina.' Some came from the States, some from the Caribbean Defense Command, and some fresh from island campaigns in the Pacific.'

"Those who had volunteered from the Caribbean met those from Stateside training camps in San Francisco and the formation of two battalions began. They sailed from New Caledonia on September 21, 1943, aboard the S. S. Lurline. Volunteers numbering 650 came aboard there, and the Lurline proceeded to Brisbane, Australia, where volunteers from the South Pacific Theater joined them. Organization and preparation continued aboard ship, and three battalions disembarked at Bombay, India, on October 31st.'

"Designated by the code name of "Galahad," technically known as the 5307th Composite unit (Provisional), they were organized in India as a single regiment after the British column system, with three columns, or combat teams, in each battalion. Later they became famous as "Merrill's Marauders," under the command of Brig. Gen. Frank A. Merrill.'

"When the time was ripe, Chinese, British and Merrill's Marauders started the push east. Familiar forces, such as Gen. Ord Wingate's "Raiders" and "Chindits," Col. Philip Cochrane's "Air Commandos," and Kachin "Rangers" under American leadership, had already won fame for their behind-the lines activities in the dense Burma jungles in the prelude to this campaign.'

"By December, 1943, the Chinese 22nd and 38th Divisions, American trained and equipped at Ramgarh, had driven Jap patrols from the Patkai Mountains, the tailbone of the Himalayas that separates the Indian province of Assam from Burma. In January, 1944, they were in the Hukawng Valley, over 100 miles from their starting point at Ledo, and were cutting into the Japanese defenses to score the first major defeats on the enemy.'

"The Marauders saw action in the Hukawng and Mogaung Valleys, and , by the end of April, they were within striking distance of their final objective, Myitkyina.'

"Myitkyina was an important objective for many reasons. It was the northern terminus of a railroad from Rangoon. It was the head of river navigation on the Irrawaddy. A little over two miles from the city was the only all-weather airstrip in this part of Burma. On the Japs' side of the ledger, it was important because it was their supply and operations base for North Burma. It was the principal air base from which the Japs had harassed American transport planes flying to and from China over the "Hump". On our side of the ledger, it was on the logical route of the proposed Ledo Road and could be our base for future operations with facilities allowing cargo and transport planes to bring in sorely needed manpower and supplies too heavy to drop by parachute.'

"After 500 miles of fighting Japs and jungles before the drive on Myitkyina began, a reshuffling and augmentation of strength of the remaining Marauders was necessary in order to have teams approaching combat organization and strength."

After bringing in replacements and assigning engineer units to fight as infantry and with help of Chinese and British troops with Kachins running interference and providing scouting, the airfield was taken. Only about 200 of the original contingent of Marauders remained for the final assault that captured Myitkyina on August 3rd. (Marsmen in Burma)

The Kachins were native to Northern Burma and had a long association with the British, having fought them during the colonization of this part of the world and eventually becoming allies. The British and Kachins formed a mutual respect from their fierce battles, both showing the other courage and ability to fight. A history from records of the Mars Task Force tells of their formation into a unit vital to the Allies. The unit was formed by the Office of Strategic Services, a highly secretive intelligence unit. Excerpts from that history follow.

"Detachment 101, is a quasi-military organization established in the latter part of 1942 chiefly for the purpose of :

Gathering and disseminating information about the enemy and the theater of operations.

Organizing and employing guerilla forces.

The operations of Detachment 101 fall into four phases:

During the first period, from January 1943 to December 1943, the Detachment, with Headquarters in India, and forward base at Ft. Hertz, had five field groups operating in North Burma: (1) The Kaukkwe Valley, (2) the Hpungin Dung, (3) The southern Triangle, (4) The Eastern Hukawng, and (5) The Dalu Area.

These first units gathered information pertaining to Japanese dispositions, strength and movement, obtained data on roads and trails, laid the foundation and started the recruiting, training and supplying of guerilla forces. The rescue of Air Corps personnel though incidental, was an important feature of the Detachment's work in North Burma.

All information emanating from 101 sources was made available to the Army ground and Air Forces through direct radio contact with the field groups. This was practically the only information from behind enemy lines and formed the basis, during this period, for most of the targets selected by the Army Air Corps.

During the second period, December 1943 to August 1944, a significant development was the organizing and equipping of over 2,500 Kachin Rangers who were used to ambush the enemy along his main lines of communication, to harass him in his retreat in the Hukawng Valley, to divert the enemy in the attack on Myitkyina, by attacks on Waingmaw and Washawng, east of Myitkyina, and to guide and form an intelligence screen for the forward Allied troops. Detachment 101 Kachins led Merrill's Marauders into the successful attack and occupations of the Myitkyina air strip and served as guides and informants to the Chinese in their conquests of Moguang and Kamaing. Operating groups with radio from Detachment 101 served as an information screen with the Wingate expedition and remained to organize resistance groups after the departure of the unit.

The third period, from the fall of Myitkyina Ausust 3, 1944 to the middle of November, was marked by organizing of the Homalin, Katha and Bhamo areas, and the reaching of the 6,000 mark in trained armed Kachin Rangers operating against the Japanese from Chindwin River to the China border.

From its inception Detachment 101 has grown from a handful of 20 American and 200 Burma personnel to 400 Americans and 6,000 Burmans, most of whom are Kachins. Among its contributions to the war effort may be listed over 2,500 casualties inflicted on the Japanese killed alone, six enemy trains wrecked, a large number of bridges destroyed, and indirectly, the incalculable destruction of enemy equipment and life resulting from Air Corps action against targets of 101 origin. 215 Air Corps personnel have been rescued from the jungles of Burma by the Detachment's field groups, and 150 wounded British and American troops evacuated by its light land and water planes. Finally, mention must be made of the inestimable good will created in Burma by the Detachment's fine American leaders and its loyal and disciplined personnel.

Detachment 101 is now entering the fourth period during which it will extend the area of its operation to Mandalay."

W. R. PEERS,
Lt. Col. INF.,
Commanding
(NACP, R.G. 407)

Kachin Rangers would be assigned from this detachment to The Mars Task Force to serve with distinction. While waiting for all the components of the Mars Task Force to be pulled together and equipped, the F Troopers had some leisure time. On a Sunday afternoon, Jack and Curtis were in the Troop command tent killing time and writing letters home. Curtis was at the typewriter, Jack at his desk.

"Murphy, who you writin?" Jack asked.

"Oh, I'm trying to type a letter to Joyce on this damn thing. It's a slow go. I wish we could have had a typing class at Garner.

"Yeah, me too. Maybe the kids will be able to have one at Millsap. That would be a good deal if you ask me."

"Uh huh. I hate to admit it, but they'll be able to get a better education there. Garner was just too small and pore to offer much." Curtis responded.

"It's a wonder we made it through Weatherford College, but we had pretty good math and English teachers," added Jack.

"Don't forget we had something else; Mamma's and Daddy's brains and encouragement."

"Now you're bragging."

"Maybe so, but that had to be a part of how we got from buck-assed privates to running this troop. It sure as hell wasn't because we're purty."

Sergeant Bill McQuary, another Mineral Wells boy had entered the tent during this exchange. "Don't you guys know why you've been put in charge of this outfit. It's not because you're smart or pretty. It's because most of us would rather go bear huntin' with a switch than to take on either one of you. Put a Jap in front of you and he's dead meat, and the brass knows it."

Jack replied, "At least you're right about the Jap. We're gonna mow 'em down."

"Horse Shit!" said Curtis smiling.

In deference to modesty they had to deny this, although they knew that there was some truth in what he said. Their work on the farm was a reflection of benevolent discipline from Mamma and Daddy and an innate sense of knowing what was the right thing to do. Duty was not an imposed concept to them, but an intrinsic , instilled response to life. Not doing their duty would be as unthinkable as spitting in a water well.

There were risks for all troopers, including the new recruits acquired to fill out the manpower shortfall. But leadership attaches even more dangerous risks. I discovered at an early age that both following and leading could lead to heartache and pain or embarrassment.

We lived on the Rawlings Place near the creek above Mineral Wells Lake. I was four years old and thought I should go with my brothers on their adventures. On a Sunday when we were all at home and the Howards had come to visit, all my brothers and Rowe decided to go to the creek. I decided to tag along. Of course, I was at the tail end of the column as we walked the trail to the creek. I think my first complete sentence was 'Wait for me!"

Unknown to me, there was a wasp nest hanging on a low tree limb just above the trail. The older boys ducked and ran under the limb, but the last one ahead of me jerked the limb. I doubt that he knew I was there because I had to stay real quiet to avoid being chased back to the house. The disturbance coincided perfectly with my passage under the nest. A ball of red wasps fell on the back of my neck. The blood- curdling yell stopped the boys in their tracks. They saw my predicament and Jack quickly picked me up and carried me to the house. Daddy smeared his chew of Beech Nut on the spot where twelve of the wasps had embedded their stingers. Did this experience teach me not to follow? Nothing could keep me from trying to follow and emulate my big brothers.

My earliest experience with leadership brought another kind of pain. Mamma always had at least one turkey hen to raise young turkeys to feed out for sale. A couple of years after the wasp incident we had a turkey hen. Mamma knew she was laying her eggs in a grove of post oaks and underbrush near the house on the Holder Place, where we now lived. She and I stalked the wily bird one day until we spotted her taking her seat on the nest.

We waited until she had laid her egg, covered it with leaves and left the area before we checked it out. She had at least ten eggs in the nest. Mamma poked around with a long stick and then covered the eggs again with leaves. I don't remember precisely why we needed to know the location of the nest, but it was always good sport to look.

I prized my possession of the nest knowledge. On another Sunday with kinfolks visiting, I decided to take a leadership role and select an inner circle to share my knowledge. With a small band of cousins and June I was ready.

"O.K ya'll listen. I'm gonna show you where our turkey nest is. Come on, I'll lead the way!"

This authoritative commotion attracted the attention of the older brothers and cousins who looked our way. Immediately a group chortle went up. They were laughing at me, but why? I had done nothing but take charge of a situation that needed direction. I quickly discovered that it wasn't my commands, but my bottom that had drawn the mirth. As will happen, circumstances conspired to bring me to a state of ridicule. We had been sliding down the cellar door. Unaware of the damage that a rusty spot in the metal door can do to a thin pair of overalls, I had worn holes in both sides of my pants to expose both my little pink cheeks. I didn't have underwear in the warm months. Henceforth, I was forever teased about my 'lead-the-way britches'. Ridicule can be as painful as the sting of wasps. One must have a strong ego to assume leadership positions.

With the same hardscrabble life on a farm, Sergeant McQuary shared these same character traits, enabling him to recognize and appreciate it in his fellow warriors. All three would soon prove that these were not empty and idle comments. Curtis gave up on the typing and stood up.

"John, uh, Lieutenant Knight, I'm goin' down to the river and check on the troops."

The machine gun platoon had been taking target practice, shooting across the Irrawaddy River that ran adjacent to Camp Landis. It was late fall and as in North Texas, flocks of geese were migrating south. In their great spearhead formations, hundreds were flying down near the river, looking for a place late in the day to feed and rest for the night.

Curtis watched this for a while. Overcome with the nostalgia of hunting ducks on Mineral Wells Lake and farm ponds and, aggravated at having no shotgun at hand, the frustrated Curtis took up one of the .30 caliber light machine guns and ordered the troopers behind him. Remember, he had had training for the action he was about to take. He fired from the hip into a large flock of the low-flying geese. Much to his surprise, one of the geese broke formation and plummeted frantically into the river. The bird flapped a few times, but couldn't become airborne. The swift current carried it rapidly downstream, finally to be submerged and become food for the hordes of catfish that inhabited the river.

Unfortunately, his marksmanship was observed by an officer from squadron headquarters and was relayed to the regimental commander. This kind of impulsive behavior couldn't be ignored by a notoriously well-disciplined cavalry unit. Reluctantly the message was passed through the chain-of-command to Jack to do something about it.

How could he deal with this? You don't put your first sergeant on K.P. You don't cuss out your brother, not if your mamma and daddy were Mamma and Daddy. Jack called Curtis into the tent. They were alone. Curtis stood at embarrassed attention. He knew that he had royally screwed up and he didn't have to be told not to do it again. Jack knew that he knew he had made a serious mistake and that he wouldn't do it again with or without comment from him. They exchanged looks. Jack was both irritated and amused, Curtis contrite.

"Sergeant, consider yourself reprimanded." That was that.

Jack's letter of December 10 contained many elements of his life at the moment: the regard for his brothers, his men, his family, the influences of his rural upbringing and the Japs. After telling about his and Curtis's new positions, he wrote, "I got two letters from you all yesterday. The last one written on Nov. 25. Boy these letters really help. I just don't have time to write much but you all know that everything is o.k. As long as you don't hear from us.'

"Oh yeah Mom please send me some pecans & some double edged razor blades. I would also like for you to send some stuff like canned chilli, sardines, & mayones, however you spell it & just anything good to eat that will keep. We don't get too much extra stuff like that.'

"I don't feel too good tonight. I guess it is because I wasted a chance today to kill a deer. I first saw the glimpse of one & didn't have time to shoot, then I heard another one barking. You see we have small about 75 lb. deer here that bark almost like a dog and I never could slip up on him. I really got mad at him. He would just stay a little bit out of sight from me. There's lots of them here. There is also a black tail deer here that weighs about 300 lbs. Some men in the Regt. have killed them. Some of my men went out yesterday & saw an elephant & a lot of monkeys & apes. Theres some pretty good jungles around here, but it's not much worse than the thickets around home.'

"I heard from Herchel. He seems to be doing fine. Fighting a lot. Well I haven't had the pleasure yet. Buzz hardly ever writes anymore. He seems kinda disgusted now. I wish I had him here with us. He would really help me out.'

"Well the war seems to be marking time right now. I hope something big breaks soon. These Japs in Burma don't have a dog's chance."

On the date of this letter, the Battle of the Bulge began.

CHAPTER 18

THE MARS TASK FORCE ON THE MOVE, WITH MULES

BOTH THE 475TH INFANTRY and the 124th Cavalry (Special) were organized for special service as long-range penetration units. The 475th was authorized 3049 men plus 208 for Brigade headquarters. The 124th, in the cavalry tradition, was authorized 2073 men which was 530 more than it would have had under regular horse cavalry organization. With a rapidly moving horse cavalry unit, the smaller unit expedited its mission. The extra men were made available because the 124th had been authorized an additional squadron for this special mission.

Each regiment had a headquarters unit plus three combat components. The infantry components were designated as 'battalions'. The equivalent units for cavalry retained its name of 'squadron.' The battalions were again divided into 'companies' and the squadrons into 'troops'. Individual cavalrymen were known as 'troopers', not soldiers. Each of the battalions and squadrons had an artillery battery attached to it. The artillery 'battery' was equivalent to a company or troop. A platoon of Kachin Rangers was attached to each battalion or squadron. The 1st Chinese Regiment had also been assigned to the Mars Brigade, or as it became known, The Mars Task Force.

The Mars Task Force was equivalent to a small division. That it could be organized, assembled, trained, equipped and supplied in four months was an incredible feat. The 475th with the remaining Merrill's Marauders moved out on November 15, 1944. They were headed for Tonkwa to the southwest to remove a relatively small pocket of Japanese occupiers that could offer resistance to the advancing British forces from the west. The 124th remained at Camp Landis for another month before they marched south to converge with the 475th for a push to the east. (Marsmen in Burma)

The last surviving letters that Mamma wrote to Jack picked up again on December 1, 1944.

"Dearest Jack,

"How goes it with you by now? We are well & hope you the same. We rec'd your letter day be 4 yest. Was so glad to hear from you again. It was written the 15th & we got it the 28th.'

"We had a letter from Loyd too. He sent 3 snap shots of himself & 2 more guys. We haven't heard from Herchel lately. I guess he has been pretty busy. "We are having some pretty cold weather now. Had ice the last 2 or 3 morn.'

"I sure don't have much news to write. We just get up & send the kids to school & Dad & R.C. go to gather pecans. I went yest. Joyce, Glenda & I just stay here & sew & play with Glenda & laugh at her. We was eating dinner today & she kinda caughed & just grunted & I said did it hurt that bad & she said uh hu. So Joyce & I liked to died laughing. She said it as plain as I ever heard any one say it. So she got scared at us laughing & went to crying. She was on the bench, so grannie had to take her. I guess we will have her talking before long. She will try to say a lot of words.'

"Jr. was surprised today. He took his tests & made 80 on algebra. He sure did work hard the last week, he and Joyce too. She helps him a lot.'

"I don't understand why we don't get Curtises letters. We haven't got but 2 since he got to where you are. Joyce has rec'd some, but not regular. I guess he is pretty busy now tho. Do you all get all our letters. I write 2 each week to all you boys.'

"We haven't seen Howards lately. Rowe was in France I believe the last I heard.'

"Aunt Fleeta wrote that Tuffy was going in the Merchant Marines in a few days. [Aunt Fleta and Uncle Dewey had a Tuffy also.]

"Dad still hasn't done anything about getting a place or R.C. a job yet.

"Looks like they will just let him go on, but I don't want him to. I hate for R.C. not to go on to school. But he isn't interested & can't stay out to go to school so he is helping in the pecans this week. They are trying to get 100 lbs. this week. We have some folks helping. Just one woman and some kids, Mrs. West. There was 2 women here this eve. Said they would be here in the morn.'

Well don't work too hard & be sweet.

We love you.

Mother & all."

June added her letter to Mamma's.

"Dear Jack,

What are you doing. I am fine and I hope you are fine too. We had our test today. She had not gave us our paper yet, only our English papers that we had yesterday and I made 95 in English. Glenda is going to sleep now. She has been talking all day. This eving Joyce had her outside. Mama was making her dress. Mama said Glenda come and help me make your little dress, and Glenda said no no. I bet that sound sweet.'

Love June

P.S. Mama and dady haven't found a pino yet. I sure hop they do. I thank you very much." [June was eight and in the third grade.]

Mamma wrote again on December 11,

"We rec'd your letter today. Was so glad to hear from you again & you were feeling well or you wrote like you felt o.k. It was the letter you sent the money order in. We thank you very much, tho you shouldn't send so much. I think the kids are wanting to go to town tomorrow to spend theirs but they won't for they went Sat.'

"Joyce & the kids went to the woods yest & got a Xmas tree & fixed it up & decorated it & have a lot of presents around it. Glenda tries to get into them, but she is pretty nice about it.'

"Glenda is looking at the fire now saying hot. It is sparkling & she don't know what to think. I don't know what we will do if Joyce goes back down to Poteet. Her mother is wanting them to come back. But I don't think I will let them go, tho I guess I can't help it if she takes a notion to go.'

"I wish you could see Glenda right now. Joyce & June was singing a Xmas song & she went to singing with her head up like she was trying to show out.'

"I will look around for June a piano when I go to town, but if it is any trouble to you, just forget it. We don't plan on anything much till you boys get home. Xmas would be mighty dull if Joyce & Glenda wasn't here.

"We haven't had a letter from Herchel in about 3 or 4 weeks. Sure would like to hear from him, but mabe he will soon. We had a letter from Loyd last Fri. He was o.k., had moved but was on the same island.'

"I hope you & Curtis gets to stay together. I know it's a help to both of you.'

"Write & tell us what your new job is. We want to know what you are doing."

June added this note,

"...Have you seen any Japs since you have been over in Burma. Are you still going to send me a necklace made out of Jap teeth.'

"Thank you for the 5 dollars. Now I can get my Christmas things. At school Mrs. Harris is letting us take time abought being teachers pet a week at a time and I am teachers pet. We are going to put on a program.'

"A while ago, I was smelling of my finger nale polish and got some on my nose.'

Love June" Joyce added her letter,

"Dearest John,

"How are things going with you? Everyone here is fine and these kids are 'bout to have a fit for Xmas to hurry and get here.'

"We got your "Xmas letter" yesterday. Sure do thank you for mine and Glenda's gift. I don't know what to do with it. I might buy her a nice doll with it and put my part in the bank.'

"She weighs 23 lbs. now (with her clothes on), so she'll probably weigh 22 pounds not counting clothes. She tries to talk all the time & copies everything anybody does. She can walk when she's holding to one of somebody's fingers, so it won't be long until she can walk by herself. I'll sure be glad to 'cause she gets dirty as a little pig now.'

"We are really getting frisky weather here now. This morning there's a lot of frost on the ground. Looks like snow, it's all over everything.'

"Well, I'll stop for this time. Be sweet and write often.'

"Thanks again for the Christmas.'

Love, Joyce

The next day Jack wrote from Camp Landis expressing his growing respect and love for his troops,

"How are things at home? Murphy and I are sitting in our orderly tent together. He's trying to type a letter but I haven't got time to fool around with that. Well another good week has passed. I'm really doing o.k. on my new job. I think I'll bring everyone of my boys back o.k. I'm giving them hell now, but I know they like it. I saw a heck of a good show tonight, 'Saratoga Trunk.' I really enjoyed it a lot. The shows haven't been too good lately."

He wrote another short letter on December 16,

"How is it all going? We are all going strong & will be the next time you hear from us. As usual, nothing is happening. This is Sunday morning & I think I'll go deer hunting again, & get one. We saw a good show last night. We have been lucky about that so far.'

"Say I'm sending my extra money home finally. I still have 280 rupees oweing to me, but I thought if I didn't send it I would be loaning out what I have now. You all can buy June a piano with it & get some bonds if there's any left. I'll send it in 3 letters..$100, 100 & 70. So let me know as soon as you get it so I can throw the stubbs away. Did you get the $40. I sent two weeks ago."

Jack's advise had taken on a more realistic and charitable tone. R. C. was anguishing about the decision to stay in a distasteful situation in school or to join the army. This is Jack's advice in his letter of December 17, the day before the march began for F Troop.

"Jew you do what you think is best. Don't let anybody influence you. If you are old enough to go to war you are sure as hell old enough to know & say what you want to do. I'm not saying go or stay. I do care wheather you go or not but it's none of my business."

Considering his high regard for education, it had to be difficult to make this concession. Perhaps he thought that military service might help R. C. change his attitude toward school. It didn't.

Camp Landis had provided a brief respite from traveling even though training and conditioning had continued. More mules had been acquired and more men gleaned from other units in the CBI. The day finally arrived.

The Second Squadron under command of Major George B. Jordon began moving out on Tuesday, December 18 following the First Squadron and 613th Field Artillery. The trail following the Ledo Road was dry and dusty. The men had spent the previous days packing their personal gear into their individual packs and assisting with assembling the mule packs. Jack and Curtis stood in front of F Troop. Jack spoke,

"Troopers! This is it. You've been told what's ahead of you. This is going to be tough, but we will do it because we must do it. If this is the only way we can get to the Japs, then that's just the way it is. Remember who you are and what you have done already. Troop! Left Face! Route step. March!"

Curtis, standing nearby, observing this movement of men he had grown to know like brothers spoke,

"John, look at those packs. Some of these guys are carrying enough junk for two."

"Yeah. It won't take long for them to learn what they need," Jack answered as he stepped to join the spirited troopers on this new adventure. They had had marches and field problems, but none with the anticipated activities at the end of this trip.

The outfit was military in appearance and conduct, but on closer inspection, uniformity in dress had been somewhat compromised. Most notable was the difference in footwear. Most of the troopers were wearing their old cavalry boots with a few inches cut off the top. Others wore infantry shoes and canvass leggings. A few had been able to purchase the new style combat boot and even fewer had the waterproof jungle boot. Appearance quickly became inconsequential. The only thing that mattered was how they fit the foot.

The first day's march took them across the Irrawaddy River and into the first night bivouac. There was little complaining. The men were determined to make this march without undue bitching. There were a few blisters but for the most part no serious problems developed. (Marsmen in Burma)

Quoting from John Randolph in Marsmen in Burma, "Not much was heard of the Mars Task Force while it was in action. There are two excellent reasons. War correspondents were kept away by stories of hardships told by rear echelon brass, and they also had the desire to get full coverage on the final opening of the Stilwell Road, which, after all, was the prime objective of the North Burma Campaign."

As Troop F began their second day on the march, the U.S.S. Maryland was anchored in Pearl Harbor for repairs of the damage suffered from the Kamikaze at Leyte Gulf on October 25. (NACP, R.G. 24)

Even though Cowboy was from West Texas and was accustomed to sand storms, this constant, clinging, pervasive dust was another reason to take exception to his forced predicament.

"Damn! I wish it would rain. I druther walk in that Ramgarh mud than breath this dust all day. If I put up with this a few days, my lungs are goin' to stop up."

"Quit bitching, Cowboy. My nose is already stopped up. I'm breathing through my mouth. My guts gonna fill up. If I had a gizzard it wouldn't be so bad, but C rations and mud really don't mix too well."

"Oh well, we just live once anyway. Gimme a light. At least we get a good supply of these nails. I know they're not good for me, but right now, I don't give a shit."

Many of the men had developed an addiction to their 'free' cigarettes. In only a few days they would deeply regret it.

During the first two weeks, the serials were marching in relatively open, level territory. Food and supplies could have been transported to them by truck. However, the brass logically decided that the air transport crews needed practice before entering the more topographically difficult terrain. Their food supplies along with mule feed were dropped from the workhorse aircraft of World War II, the C-47. The practice paid off as the pilots and crews developed coordination for fairly accurate drops.

The system worked well as long as the crews had good visibility. (Marsmen in Burma)

After several days out and on bivouac at mid-afternoon, Curtis and Billy Ralph Aaron decided to go deer hunting. They had become reasonably adept at hunting Texas Whitetail deer during their maneuvers in south Texas. At this time, there were few deer in Parker County. They were being re-established and it was illegal to hunt them.

Daddy made the most of this situation in the 1930s when we lived on the Crosthwaite Place north of Garner. Immediately west of the farm was a dense growth of post oaks and blackjack oaks with heavy underbrush. We called it the 'Thicket'. The Thicket made excellent cover for the deer. Daddy planted a sizable crop of peas for maturity in early fall. Any kind of pea, black-eyed, purple hull or cream is an irresistible feast for Whitetail.

Daddy would go before sunup and hide in the adjoining corn rows. At the first hint of light and with good luck, he would use his new full-choke 12 gauge shotgun with double ought buckshot to kill one of the deer. He had the older boys waiting at the barn with the wagon and team. They would hurriedly move the wagon on hearing the shots. Daddy would quickly load and cover the deer with cotton sacks, take the animal home, dress it and prepare it for canning. Mamma would chicken-fry it in strips and use the pressure cooker to can it. Later, as the meat was used, the jars yielded tender pieces of meat with a delicious gravy.

This was another way that my Dad found to feed a family with eight children without going on the welfare roles. He took great pride in his making it through the Great Depression without going on relief, even if some of his methods were illegal.

He had little good to say about the 'New Deal.'

Curtis and Billy Ralph planned to find a good spot near an opening in the brush and wait for a deer to appear. There was a large variety of deer that could grow to 500 pounds and then there was the smaller, barking deer. They saw neither. They heard the bark of the smaller deer, but none appeared in the opening. Finally, they resorted to the old trick of walking through the brush to kick one into the open. Billy Ralph would do the walking and noise making. Curtis, because of his reputation as a marksman would be left as the shooter. They heard more barking and running.

As darkness fell, the two gave up and headed back to camp.

"Those barking little rabbits sure got our goat, didn't they?" said Curtis dejectedly.

"Yeah, they damn sure don't act like our deer at home," answered Billy Ralph.

"Maybe we ought to talk to some of the natives to find out how they do it."

"I'm all for that, if I could speak Burmese."

"I'll bet we could get one of the Kachins to interpret. Maybe they could tell us."

"Yeah, I guess."

They ate their sustaining, if not tasteful C rations and prepared for the night. The troopers had each been given a shelter half and a blanket. The S.O.P. was to find as soft a spot on the ground as possible. If available, leaves were used to cushion the earth. More comfort could be achieved by digging out a place for the hip to fit. The men could sleep alone by wrapping in the blanket and laying on top of the shelter half. They could team up to build a pup tent. This approach was more popular later as the rains came. The greatest inducement to sleep was the preceding day of exhausting marching. They would find out that, up to now, they weren't even closely acquainted with exhaustion. (Marsmen in Burma)

Mamma and June wrote on December 19, the day after Troop F started their journey through the North Burma jungles.

"Will try & write again. We have been so busy & had company too. Mary stayed from Fri. till Monday with Joyce.'

"Hope you are all o.k. We are well. We rec'd your letter about the program Sat. So we stayed up trying to find it, so at 11:30 Dad & I decided to go to bed. About time we got in bed Joyce turned on a station that we didn't even listen on & found it so we was afraid to turn on another station to find it. Afraid we would miss something. We sure got a thrill. You should have seen us. Joyce, Mary, Jr. & I were all hovered around the radio like we was afraid it would run off. We could hear you boys yelling & whisteling. We tried to pick you boys out but too many. I sure am glad you wrote us about it and your letter got here sooner than any of the rest ever did.'

"We went to see R. C. Sun. He is doing fine. We got his draft card yest. He was put in C2 so mabe he won't have to go.'

"We hear from Loyd regular, but haven't heard from Herchel in 5 weeks, tho we might any time.'

"The kids sure has had a big time shopping for Xmas. I wanted them to get something just for their selfs with the money you sent, but they just bought for each one of the folks & school kids. June got her a little necklace. It is pretty.'

"We went to look at a piano yest & it was sold. It was a good one. It was $100. I want to find one while Joyce is here so she can help decide on it & she can teach June the notes.'

"I'm in a hurry to get this off on the mail and write Loyd too.'

"Oh yes. I got me a nice Bible from you. I sure am proud of it.'

"Well be sweet & write real soon. We love you bushels-give Curtis my love.

Mother and all"

"Glenda is getting smarter ever day & so sweet. She will cry after me just like I was her mother. So I am anyway."

June's letter,

"What are you doing. I am fine and hope you are fine to. I got the 5 dolars. Thank you for it.'

"I got me a locket with a army insiginia in it. Boy it realy is pretty.'

"Joyce has 200 70 c in dimes. She is saven them. Bill is making the funniest picture I ever saw. Do you rember when I would draw picters in your letters when I couldent think of iny thing to write. I beleve I will draw smoothing. Here it is.

Love June

The 475th Infantry Regiment moved out with 3137 men, approximately 100 short. The 124th Cavalry Regiment Special was allowed 2173 men but also failed to reach full strength before their march began. The troopers would exchange USO shows with the likes of Jinks Faulkenburg and Pat O'Brien for the entertainment provided on the Japanese radio by Tokyo Rose.

The First Chinese Regiment moved out after the 475th. With great resentment from the U.S. soldiers, the Chinese were being transported by trucks, staff cars and jeeps along with amenities not available to the Americans. The fact that much of this equipment was there because of U.S. Lend Lease did not escape the eye of the men of the Mars Task Force. Overlooking the possibility that the Chinese could be serving as a diversion for the Mars Task Force, the Marsmen had plenty to say.

"Look at those lazy-assed Chinks. How is it they get to ride while we walk and drag these stubborn mules?"

"Hey, bitching won't help. Besides, I don't remember anybody telling us this was going to be easy."

"Look at it this way, they'll get there first and maybe the Japs will run out of ammo before we get there."

"Yeah, but they won't be exhausted when they get killed!"

'There' was approximately 300 miles south of Camp Landis near Namhpakka, Burma where the Japanese were strongly entrenched on a stretch of the Burma Road. Some estimated that the actual marching miles was nearer 450 miles. The Ledo Road intersected the Burma Road a short distance north of these Japanese emplacements. U.S. construction battalions were at work improving the Ledo Road south out of Myitkyina. The Marsmen were able to use this improved roadway for the first couple of days out of Camp Landis. The ease of these first days was illusory because their march quickly left this smooth, wide surface for the narrow trails of the Burmese jungle and soon the mountains on the eastern extremities of the Himalaya range. Even then, their route remained close and somewhat parallel to the Ledo Road for the next several days. (Marsmen in Burma)

As the columns trudged along the dusty trails, physical and mental fatigue set in. Days were very hot. The latitude was approximately 25 degrees North, about the same as Central Mexico. The physical fatigue was partially alleviated by ten-minute breaks every hour. Some of the mules learned to sit down to lighten the load during the breaks. They couldn't lie down because of the bundles of supplies and equipment strapped on their backs and sides, but a good sit was restful. Even though the men had received rigorous physical training at Ramgarh and Camp Landis, all day marching was a grueling task.

Reconnaissance information gave the troops reasonable assurance of safety from Japanese attack at the early stages of the march. Mental activity wasn't concentrated on vigilance but on thoughts of home, the past times at Ringgold and the certain contact with the enemy a few weeks away.

I was eleven years old December 23. Jack wrote on that day,

"Had a little time & thought I'd write. Everything is going fine. Pot Guts & I are still together & having a good time. This is a cinch for most of these thugs of mine. I really enjoy being with all these boys in F Troop.'

"Well we have some good hunting ground tonight. I can just taste that deer steak.'

"We sure are lucky not having any rain these days. I'd hate mudy weather."

On the same day General George Patton's Third Army relieved Bastogne to free the 101st Airborne and turn the tide on the Germans for the last time. The advance of the Allies was virtually unimpeded after that. (WWII, American Heritage)

During that winter of despair at home about the only relief was Glenda, the cherished first grandchild. Mamma wrote to Jack the same day, December 23,

"Glenda & I are here alone this eve. Joyce went to the Xmas tree. The kids has been after her to go to school, so she went with Mrs. West. I enjoy keeping Glenda. She's pretty good. Just plays around in the Xmas presents now, but she hasn't tore into any of them yet. I just got her puppie in the house & she is wooling him now. He's black. She tries to lift him.'

"The kids had a little Xmas prog. last night & we went. Bill & June were in it. Bill is in the glee club. He & June both sing & June spoke 2 little pieces in a little play.'

"I guess we will finish up the pecans next week or we should. Then I guess Dad will get a job in the camp. We had a letter from Loyd yest. He was ok. & we have had several from Curtis. Joyce got several letters that was written about the middle of Nov.'

"Jim Kidwells boy is home from the South Pacific. He said he had seen the Maryland, Herchels ship, but didn't remember when. He said it had made a trip to the states to transport soldiers, so they may be on a trip now.'

"I hope you boys are doing ok. & not too much to do, tho I guess it will be ok. As long as you don't have to fight.'

"Well this is after supper. We are just sitting around. Jr. & Joyce is matching pennies. Bill & June watching them. Dad listening at a radio play. Glenda has gone to sleep, something unusual. She generally stays awake till all the rest are in bed.'

"I don't know what we will do for Xmas. We may go to Grannies awhile I don't know. R.C. will be home Xmas. Don't know just when he will come home. We thought mabe Sat. night.'

"We are making some pictures now. We'll send you boys some soon as we get them developed.'

"Well I just about went to sleep listening at Gabrel Heater give the news report, tho I guess it's because I got still. My mind wasn't asleep. The news in Europ isn't so good now, but I think it will turn for the better pretty soon. I sure hope so.'

"The kids just get one week for Xmas, but I guess that's long enough. One of the kids got some clay for Xmas & they've been making noses and put on their face."

June added a note,

"What are you doing this fine morning. It is Saterday, Bills birthday and they are trying to make him work.'

"Glenda sneezed and wiped her nose. She wants down so much. Joyce is trying to put on her shoes. Yesterday we had our Christmas tree and I got 5 preasetants. A come and a come with a riben on it and a puzzle and some clay and braclet."

Love June

Mamma added a P.S. "We was made happy this AM. We had a letter from Herchel & he was at Hiawaii, so we are so glad, tho we haven't had a letter from you since last Sat. but that's not so long. Joyce had a letter from Curtis & that is like hearing from both of you. Excuse this writing. Glenda caused it. Ho ho."

Love Mother

Troop F continued their march near the Ledo Road. The terrain was not difficult for them and they had time for some hunting and a break to observe Christmas. Jack wrote on Christmas day,

"Merry Christmas Gang,

"How is it going? Sure wish I could be with you all today but I guess I can be thankfull that I'm resting with a bellie full of good US rations. Some of the boys are tanking up on Sockie & Chinese whiskey but I can't go for that. They can swap the natives out of it.'

"Curtis & Billie R. went out deer hunting last night. They saw one but they are so wild you have to be quick on the draw. They didn't get him.'

"Well I guess we just have to wait for mail. We hardly ever get any. I got your box about a week ago.'

"Well I can't tell you anything now yet but maybe I can later. We are all ok. Working hard & in good health.'

"Be good & write often. Tell me if you get the money I sent."

Love, Jack

Mamma and Joyce wrote to Jack the day after Christmas,

"...We are all well and as mean as ever. The weather is so bad we have to stay in and this is the kids week off from school.'

"Since I started this letter we've been laughing at Glenda. She got down in the floor & began crawling & turning over & around just to make us laugh. She is the smartest kid I ever saw.'

"Everthing is covered in ice now. It sure looks pretty, but not very pleasant outside.'

"R. C. came home sun. night & stayed till last night. He is doing fine at his job, gets up at 4:30 each morn. They milk 58 cows. They use milkers but have to strip them.'

"We heard another program last night from Burma. It was 219th U.S. Airforce bombing the last bridge in the Burma Road. The man in the plane was telling just what they were doing & about the enemy planes coming up to meet them.'

"We heard today on the news that the Japs were puzzeled about the strength of our air force over Burma & India. That makes me feel good. Everthing sound good now everwhere.'

"We had another letter from Herchel today, written Dec. 8th. It was before the other one was written. We had one from Loyd too. He was ok. Not too much to do, he said. He wished he was with you boys but said he was satisfied anywhere now. Said he had a letter from you.'

"We went to Grannies Xmas day. Aunt Rosas family were there. Kenneth came home on a 3 day pass. We got to see him. He went back yest. He's leaving Camp Polk today for a camp in Kans. He may be leaving soon. I hope he doesn't have to go across for he will be a mechanic on a plane.'

"Grannie & Sam is ok. Just like they have been the last 3 or 4 Years.'

"I wanted to go see Howards this week & Aunt Matties too, but I guess the weather will be too bad. Wade told Dad Rowe is in Belgum now and Norris address is Luxinburg & Freddy was in France. So we are anxious to hear they are o.k.'

"Well this is later. I haven't much news so I had better close, hoping to hear from you real soon.'

"All our love & prayers are with you all.'

Mother Dad & all."

"P.S. Jack your things from India or wherever you sent them from hasn't got here yet & so far we haven't got any of your bonds."

Excerpts from Joyce's letter,

"We got a letter from Loyd and Hershel today. Loyd said he was getting along oke. Jr. sent him an unsolved algebra problem & Buzz solved it, so Jr. lost 75 c. over it. Those two are just about alike.' Buzz said he had heard from you twice and you was telling him about all the old men in the troop. He said he sure would like to be back in the troop, but he was satisfied where he was. He was telling us all about you being the troop C.O. & called Curtis "Dog Murphy". Some of these days I'll get around to seeing all the Gerts & Murphys and Bill N...... & the R.. H...... & the J..... C......'s I'll have to look over this Parker County good when Curtis comes home. (Let's hope it won't be long 'till you are all home.)'

"We all went to Granny K's for Xmas dinner. We ate like a bunch of "horgs" but it was swell. After we ate, Smutt , Bobby Armstrong, Marjorie A., June, Bill & I went to see Abbott & Costello in "Lost In Harem." It was a good one."

The 124th Cavalry Regiment came to the end of the 'easy' part of their march shortly after Christmas. Two weeks into the march orders were received to change directions, abandon the route near the Ledo Road and proceed through a much more mountainous terrain to the Sweli River in the vicinity of Mong Wi. On topographical maps, if the Himalayas were a dragon, this range called Gaoligon Shaw, would be the end of the tail. Maps showed trails suitable only for mules and men. No type of motor vehicle could make it along those trails. At the last bivouac before heading into the mountains a three-day supply of food was received by air drop. A few replacement troops joined the 124th. Over the next few weeks they would depend upon inconsistent air drops and their own creative foraging for their food supply. The troopers learned to trade unwanted government issue cigarettes and unneeded rupees to the Burmese for such delicacies as chicken, eggs and vegetables.

In early January, 1945 Daddy received a letter from the 124th Cavalry Headquarters,

"Dear Mr. Knight,

"Lt. Jack L. Knight IS NOW PARTICIPATING IN MILITARY ACTIVITIES OF THIS UNIT WHICH WILL PREVENT HIS WRITING TO YOU FOR A TIME. DURING THIS PERIOD THIS HEADQUARTERS WILL INFORM YOU TWICE MONTHLY AS TO HIS WELFARE.'

"WHILE HE WILL BE UNABLE TO WRITE TO YOU, IT WILL AT TIMES BE POSSIBLE FOR US TO DELIVER MAIL TO HIM. MAY I SUGGEST YOU WRITE HIM OFTEN, AS I KNOW YOUR LETERS MEAN A GREAT DEAL TO HIM.'

"AS YOU KNOW, MILITARY RESTRICTIONS ESTABLISHED FOR THE SECURITY OF THE TROOPS, PREVENT US FROM GIVING YOU MORE DETAILED REPORTS. I KNOW YOU WILL UNDERSTAND THAT I WILL NOT BE ABLE TO REPLY DIRECTLY TO LETTERS REQUESTING ADDITIONAL INFORMATION, AS ANY SUCH INFORMATION, WHICH THE AUTHORITIES ARE WILLING TO RELEASE MUST COME TO YOU THROUGH REGULAR ARMY CHANNELS.'

"WHEN ADDRESSING HIM PLEASE MAKE CERTAIN TO INCLUDE HIS FULL NAME, ARMY SERIAL NUMBER, TROOP, AND UNIT DESIGNATION, HIS APO IS #218, c/o POSTMASTER, NEW YORK, N.Y.

FOR THE COMMANDING OFFICER:
R. B. SATTERLEE, Jr.
CWO, USA
ASSISTANT ADJUTANT"

All vehicles were worthless and had to be turned in at Momauk. The much-heralded jeep that could go anywhere couldn't traverse a three-foot path on the side of a sixty degree mountain slope. Even the carts had to be abandoned during the first day on the mountains. Much to the dismay of the troopers and the life of the mules these conditions proved too much for many of the animals. Too many lost their footing and plummeted to their deaths with the loss of not only the mule, but the precious supplies they carried. Some supplies and mules could be rescued, but others were simply left in deference to the dangers of rescue and retrieval. (Marsmen in Burma)

Two weeks on the road with hills and dust before reaching the mountains had helped physically prepare the men for the extreme physical challenge that lay ahead. With the necessity of abandoning equipment and supplies along the way, a fourth shakedown was called for before entering the mountains. On the first day out, the path up that mountain was strewn with carts, harness, cooking gear, mess gear and various items of clothing, even ammunition. The local Burmese would discover a windfall of valuable items that would temporarily enrich their lives with more rupees and useful tools and appliances. (Marsmen in Burma)

Soon after entering the mountains heavy and persistent rain began with its cloud cover obscuring the trail. The dry dust had been choking, but the rain presented other, more serious problems. The excessive rains softened the soil along the narrow, steep paths. One misstep by man or beast could cause quick descent into the ravines below. The cloud cover made it impossible for the airplanes to find the troopers for supply drops, causing delays in food delivery of three to four days. They could usually camp on mountain tops flat enough to allow relatively safe sleeping. (Marsmen in Burma)

Troop F had two mules go down the side. One of these mules was carrying a heavy .50 caliber machine gun. A squad of men slid down the mountain to investigate. They found the mule too severely injured to continue. One of the men followed tradition in these circumstances and shot the mule with his .45 pistol. The men carried the machine gun back up the mountain, rearranged loads on the other mules and remounted the machine gun. This meant that some of the men acquired additional load. By this time every weapon and pound of supplies were too precious to be abandoned.

During the long delayed airdrops, the men subsisted on the brick hard chocolate bars and crackers that were always the last of the rations to be consumed. On the third day of one of these delays, Jack became impatient.

"Murphy, if we don't get a supply drop by tomorrow, I swear, I'm gonna kill one of these mules and barbecue him."

"You go right ahead, but I think I'll pass on that for now."

"I've always thought I was tough enough to take anything, but nobody can do without food and keep going at this pace," Jack complained.

"You're right, but we'll be supplied in a day or two. We may not be able to move over these mountains as fast as we want to, but as long as we have water and these delicacies, we'll live," teased Curtis.

Jack added, "Yes, and as long as we have these mules we won't starve."

As they moved into the higher elevations, the nights were bitterly cold. The combination of cold and hunger brought misery to the troopers. They were able to improve their circumstances in small ways. They doubled up and slept in pairs under both blankets and shelter halves to gain moderate warming.

New troopers were added along the trail, flown in by the tiny L Planes that had become indispensable to this moving brigade. One of these young men, after his first night on the trail was sitting on a log. As Sergeant McQuary walked by, he noticed big tears running down his face.

"What's wrong, trooper, you lose your best friend?"

"No. My canteen is frozen. I can't get a drink. We'll never get out of this alive."

Sergeant McQuary walked on with a notion that the kid might be right. Bill McQuary told me this story in April, 2002.

The next day one of the muleskinners came to Curtis with a problem. One of the mules had received an open wound injury and was badly infected with screwworms. The mule couldn't continue. Curtis ended his misery with a forty-five slug. With mischief on his mind he went back up the trail to find Jack. He found him with several of the men.

"Hey, Lieutenant, I just shot a mule that was down with screwworms. You want us to barbecue him?"

"Hell no! I'm not that hungry, after all."

Aside to the witnesses, "That guy has always been a smart-aleck."

Most of the month-long trek was sheer mind-numbing, energy-depleting drudgery. Wake up early, eat a meager breakfast, check the mules and loads, climb a few miles in single file with little conversation before collapsing at the end of the day for a nights sleep on the ground with not enough blankets to cover from the cold.

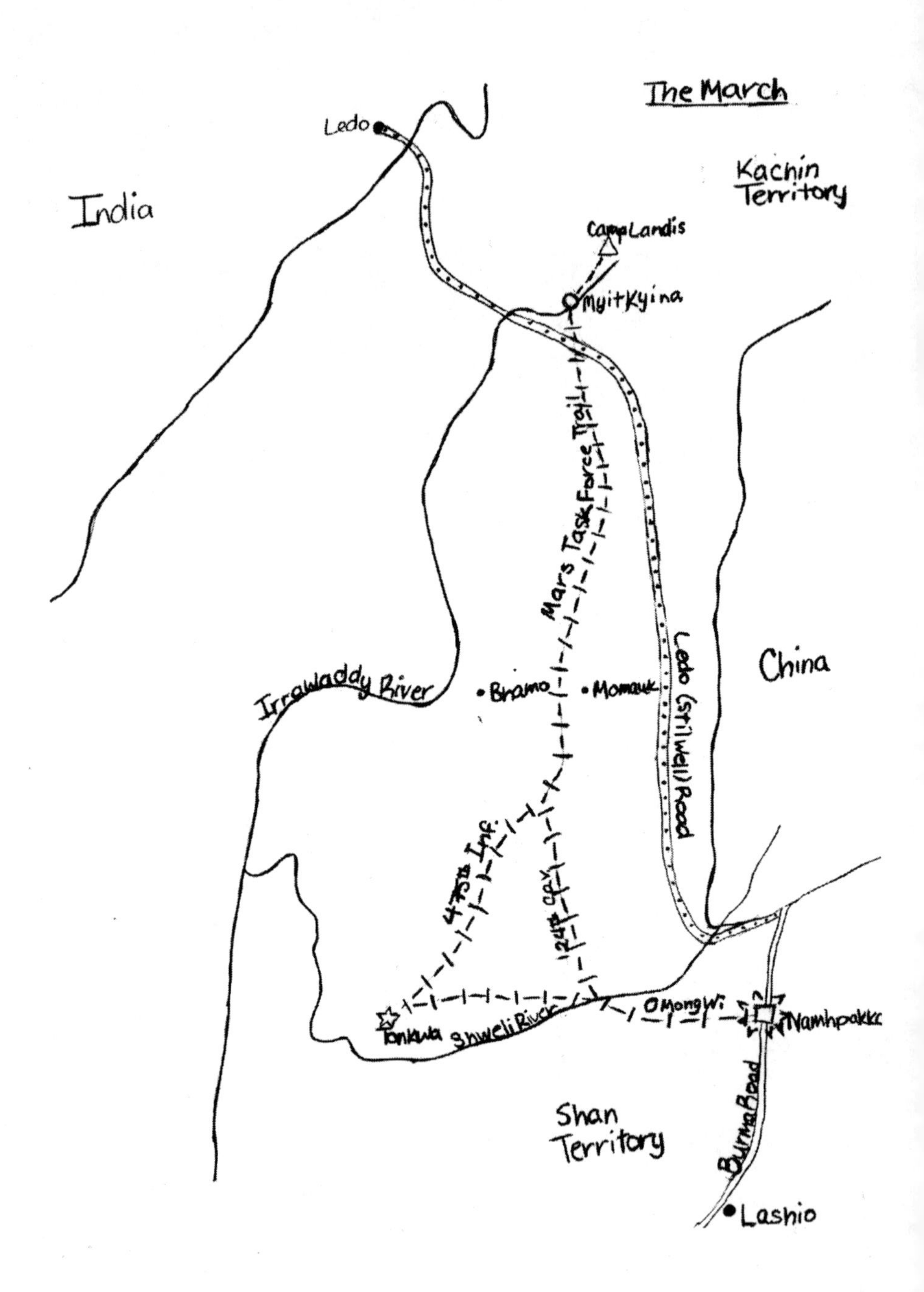

The March
Ledo
India
Kachin Territory
Camp Landis
Myitkyina
Mars Task Force Trail
Ledo (Stilwell) Road
China
Irrawaddy River
Bramo
Momauk
475th Inf.
124th Cav.
Tonkwa
Shweli River
Mong Wi
Namhpakk
Shan Territory
Burma Road
Lashio

CHAPTER 19

THE MARCH, A MONTH

THE FIRST DAYS OF 1945 saw the Allied forces ready to pour into Germany all along its western borders. The Yalta Conference was being planned for early February in which Roosevelt, Churchill and Stalin made decisions relative to the political face of Europe after the war. The Russians were poised to sweep into Germany all along the eastern German frontier.

American soldiers began the largest land battle of the Pacific war on the island of Luzon in the Philippines. Savage battles ensued, costing thousands of American lives and multiple-thousands of Japanese lives. The Filipino civilian population also suffered thousands of casualties. The Japanese were never completely eradicated until the end of the war. (WWII, American Heritage)

The 475th Infantry that had left Camp Landis on November 15 had engaged the enemy near Tonkwa on December 9. This engagement lasted through December 22 when the last Jap straggler was killed after his comrades had evacuated on December 18. The 475th had killed 220 Japanese in this campaign. They were then ordered to move east to the Burma Road on December 28 and were on the march by January 1. The 124th Cavalry entered the mountains on their march to the same destination. (Marsmen in Burma)

The march had become arduous, mind-numbing routine. But there were some interesting stories that came out of this experience. John Randolph, in his Marsmen in Burma, told of the following personal observation,

"Once I was well ahead of the column when I met two natives. Each had a chicken secure in a basket woven of bamboo strips. I had no desire to lug a chicken with me, but I tried to find out what their price was. They showed little interest in a trade, but as we talked without understanding each other, they squatted down beside the trail, put their chickens in front of them, and spread out a cloth. I gave them each a cigarette, which they carefully laid on the cloth. Confident that some GI would make a trade even while keeping up with the moving column, I stayed to watch. One man after another tried a fast bargain to no avail. The natives sat still and smiled. The two cigarettes grew into a pile of loose ones and partial packages. D Ration chocolate bars and assorted fruit drops from 10-in-1 rations were tossed on the cloth. Finally they both got up, took their chickens to the next two men to come along the column, handed them to the amazed troopers, picked up their cloth full of "pay," and walked away."

Along the trail the mules at times offered comic relief. Again quoting from John Randolph,

"Pfc. Raymond O. (Chigger) Wall, of Oklahoma City, Okla., had a mule named Mitch. Mitch, like all mules, was herd-bound, not only to mules but to men. Frequently we would come to little streams crossing the trail where there were more or less fragile footbridges. Sometimes the muleskinners would use the bridge and let the mules ford the creek. At one of these points, Mitch refused to get his feet wet, seeing that Chigger was keeping his dry. The bridge was narrow and weak, but Chigger could not keep him off. Mitch had to put one foot gingerly in front of the other as he walked slowly and safely across, while the men behind yelled at him, and Chigger visualized his slipping off or at least breaking the bridge down so that it would be useless for all the men behind him. Mitch proved his eligibility to join the circus, if they would just take that heavy load off his back and return him to civilization."

The troopers also found that some of their orientation, although given in sincerity, proved to be misinformation. Once again, we lean on Mr. Randolph.

"The same authority who had lectured to us about the beautiful women we would see in Burma also told us that it never rained there between monsoon seasons unless it rained during the week between Christmas and New Year's. He promised that, if January 1st came without rain, there would be none until summer. At 0500 on the morning of January 6th, we were awakened by rain. We were to think often of this man's second misstatement as we lived through the next few hardest and most exciting days of our march."

The 124th finally reached the Shweli River after what seemed an interminable trek through the mountains. The Shweli was west from the highest mountains and spilled into the Irrawady. An expected easy decent to the river was not to be. The rains had saturated the fine red topsoil and after the first contingents made their way down the mile-long slope to the Shweli valley, the trail became slick and treacherous. The Chinese on the mountain outpost had added to their woes by wearing a trail to the river and airdrop area. The 475th had arrived on January 7th and crossed at about the same time of the British unit. Their combined wear and tear added even more to the slide. (Marsmen in Burma)

The Second Squadron was the last to approach the river. Men of Troop F were among the last of more than five thousand American soldiers and more British plus a thousand mules to take the plunge, and a plunge it was. Jack and Curtis were the first F Troopers to peer over the edge and witness the impromptu skid.

Curtis was the first to speak,

"My God-a-Mighty! Look at that. You think you can keep your feet on that shit?"

"Well, all we can do is try. Better make sure you have some paddin' on your ass, 'cause you're goin' to spend more time on it than your feet before you get to the river," Jack replied.

Jack cautioned his troops about the conditions and waited for the mile-long slope to clear. He ordered his men over the precipice and they put on quite a show. Efforts to walk down were useless. Those with very good balance could take a step or two, slide a few yards to more secure footing and remain on their feet most of the time. Most slipped and slid down the slope on their butts with various trajectories. Some used their feet to guide the direction of their decent, while others kept their feet tucked in and spun their way down, not unlike sliding down a snowy hill back home on a piece of tin or cardboard. The mules, a more pragmatic species than man, simply sat down and let it go, guiding their descent with their front feet. A few of them lost control and tumbled until they could regain their footing to continue. After going over the edge, there was no choice but continue and finally all the men, mules and supplies reached the valley, more or less intact. The bivouac area near Mong Wi was two days away.

Somewhere along the way, the troopers performed another slapstick comedy routine with the mules. A column had halted to rest on a hillside that offered good grazing for the mules. A fallen tree in the meadow offered a good place to tie the mules with enough line to allow them a good range for grazing. Unknown to them there was a beehive hidden in the limbs of the tree. Of course, the mules in their quest for forage, disturbed the bees. These especially large bees took offense to being disturbed by these large mammals and attacked. A real rodeo ensued with the men trying to untie the mules and both trying to escape the vengeance of these fierce insects. The men survived relatively unscathed. Not so for the mules. Three of them died and another was saved by the veterinarians by a rare blood transfusion. (Marsmen in Burma)

There were some benefits to being last. The first troops of the 124th to reach the Shweli, the swiftest running river in Burma, had to complete the work of the 475th who were the first to cross.

They had found a very primitive and unstable bridge over the 400 foot width of the river. It was constructed by the Chinese from a few pontoons, bundles of bamboo and canvas, covered with dirt and grass.

The American troops had equipment and animals too heavy for the makeshift expanse. The 475th had already rebuilt the bridge using bamboo rafts, with more bamboo providing floor. To anchor the bridge, picket lines, lash ropes and jungle vines were used.

Men were assigned to make constant repairs as the crossing continued. The regiment at last managed to cross without loss of life, men or mule. With this fearful experience behind them, some would think the Japanese soldier less of a threat to their health and safety. They would go on to Mong Wi for a few days rest and then to the most difficult leg of the march over severe mountain trails.

At Mong Wi, east of the river, a drop area had been established and the air-drop re-supply had begun. All units of the Mars Task Force converged on this spot in some of the most remote and inaccessible territory on earth. The men had little concern for their exact location or the nature of their surroundings. They needed rest and food for revitalization necessary for the traumatic experiences awaiting them the next few days.

Rain continued and the camps became quagmires. The cloud cover prevented the airdrops of supplies. Airplanes could be heard overhead searching for them, but the effort was fruitless. The troops had cleared a large area for the eventual drop. An evacuation station was also established. Several men, including the Commanding officer of the 124th, Colonel Heavy, were in no shape for combat and had to be flown out for treatment elsewhere. Colonel William Osborne who had only recently recovered from an illness took command of the 124th. He had been with the 475th as an advisor. His experiences with Merrill's Marauders would help.

On the third day of the bivouac, the skies cleared. The 124th had arrived only the day before. Troop F had acquired a few new men on the march, their arrival made possible by the small single engine planes landing on very small fields in the valleys along the way. One of these men was Pvt. Guy I. McNutt. Having been with the troop only a few days, he had not mastered all the safety precautions necessary to survival in this strange world of men, mules, mountains, rain and air drops. On the tenth of January he was trying to do his part in this war against the totalitarian warmongers of the Japanese Empire. Certainly he expected to make his contribution by killing a few Japs and eventually go home to the admiration of his family, friends and country.

On this day he was sitting on his helmet near the drop area cleaning his rifle. The food rations for the troops were being dropped with parachutes. The bags of oats for the mules were being pushed out to free-fall to earth. Sometimes all of the stack did not fall out on first push. This day, instead of circling to push out the remainder, the pushers just kept kicking it out and were out of the drop zone when the last bags fell. Being eager to prepare for battle and without the precaution to look skyward as the planes flew over, he was unaware of the oat sack that hit him. He was evacuated immediately, but died the following day in a rear hospital, an inglorious yet sacrificial end to his life and expectations. That's life and that's war.

Each unit was supplied and it was time to make the final push to the battle zone. Two days over the highest and most treacherous mountains yet encountered lay before the men of the Mars Task Force before reaching the Burma Road and the Japanese armed forces. Their trails would lead them above the timberline. (Marsmen in Burma)

It's a well known adage, reinforced and expanded by Mark Twain that, "familiarity breeds contempt and children." This was certainly true for at least two men with Troop F roots. In the beginning at Ft. Bliss and early at Ringgold, all the troopers were natives of Mineral Wells and the surrounding countryside. Soon men from other parts of Texas were assigned to the troop and later still, men from all over the United States were transformed to F Troopers.

One of these men was Cowboy, with whom you are acquainted. He was not a model of traditional cavalry discipline or protocol. He liked to ride his horse as he had done on the ranch back home. Posting seemed to him to be a little silly. He liked to take his alcohol on his own schedule as I have related. He possessed the free spirit of a stereotypical Texan. Lt. Jack 'Snake' Warner, a native of Alabama, former troop commander, found these traits offensive and threatening, especially Cowboy's disdain for military courtesy. Their contempt for one another sparked at Ramgarh without injury to either, but it festered in both men throughout the march from Camp Landis. They were kept apart on the trail because of the necessity for single file movement and the fact that Lt. Warner was now assigned to Squadron Headquarters away from the immediate proximity to Troop F.

Upon the arrival at Mong Wi the Second Squadron was bivouacked in the same area and the chance meeting of the two was inevitable. Cowboy, now on the verge of combat with the Japanese and the very real possibility of death looming, didn't really give a damn whether he pleased anyone, especially Lt. Warner. He simply ignored the lieutenant and abandoned any attempt at military courtesy. Lt. Warner was incensed and because of their past animosity, decided he would again try to bring the young man in line. He sent word to Cowboy to come to his tent. Cowboy could be expected to ignore this message, but perhaps because he relished a confrontation with his adversary, he went.

"Private, you have been a pain in the ass since you were assigned to the 124th. I will not tolerate your attitude. You're still an enlisted man and I am an officer and I expect you to extend me the same courtesies that you give the other officers."

A short silence followed.

"Private, do you understand me?"

"Yeah, I understood what you said," snarled Cowboy with as much sarcasm as he could muster.

"Do you have anything to say for yourself?" shouted Lt. Warner, anger swelling.

"Go screw yourself, 'Snake'," he replied with an even glare.

Flash point anger exploded. Lt. Warner, using his skills learned in boxing and wrestling quickly landed some painful licks on the antagonist. Cowboy could handle himself in a brawl and managed to land one good punch to the lieutenant's lip. The ring on his finger split Lt. Warner's lip. Quickly, other men pulled them apart and dragged Cowboy to his tent.

One of the F Troopers ran to let Jack know what had happened.

"Lieutenant Knight, Cowboy just had a fight with "Snake' Warner."

"Shit! Let's go, private. Curtis, take over 'till I get back."

They didn't need this kind of conduct on the eve of combat. As they hurried away Jack inquired of the messenger,

"What happened, trooper?"

"Lieutenant Warner was chewing out Cowboy and Cowboy said something that made Lieutenant Warner mad and he jumped on Cowboy and beat him up, but Cowboy knocked the shit out of him once."

No one knows what ensued in the following scene except Jack and the 'Snake.' I asked Curtis and he told me that Jack never told him. In a telephone conversation with Jack Warner, now owner and CEO of Gulf States Paper Corporation, on August 17, 2000, he didn't respond when I asked him about it. Not knowing definitively, but with the help of a few grizzled Troop F veterans that I have interviewed, I can relate the following description with some confidence.

Jack was angry as he arrived on the scene. Cowboy, red faced and with anger still flashing, was being led away from the area. Lt. Warner was sitting on an ammo box, pressing his handkerchief to his cut lip, trying to stop the bleeding.

Jack headed straight for him. Warner had chosen to reprimand one of his men directly, bypassing him and creating a nasty incident that was both foolish and destructive to the camaraderie that was needed in battle. Curtis positioned himself nearby, simply as a precaution in case Jack needed him. He wasn't sure how Jack would handle this because he knew how protective Jack was when it came to his beloved F Troopers and though usually even tempered, could respond with physical anger.

Only Mr. Warner knows what transpired over the next few minutes in this conversation between two officers of equal rank. I can only assume it was most unpleasant. Perhaps it is best to let some memories die. After the short and intense confrontation, Jack stood and walked away, leaving Warner to contemplate the cost of personal grudges and vengeance. A few men of Troop F had positioned themselves to witness this confrontation. They were disappointed that they couldn't hear all that was said in the conversation, but the demeanor and body language of both men left little doubt about the message. Alvin Moberly, a private in Troop F at the time, swore to me in an interview in April, 2002 that he heard Jack threaten to cause him severe bodily injury. Curtis told me that the only thing Jack said, as he was leaving, "Curtis, we won't have any more trouble out of him."

Word quickly spread. Jack's reputation for devotion to his men was solidified. Lt. Warner had done Troop F a great favor. They proved a few days later that hell on earth couldn't keep them from following Jack.

Mamma had written on January 7, 1945,

"We went to Aunt Matties this eve. & went to Sun School & church this A.M.'

"Aunt Mattie's folks are well. They heard from Fred & John this last week but haven't heard from Norris lately.' [The Watkins cousins were fighting in Europe.]

"We sure do miss Joyce & Glenda. I wish now I had begged them to stay, but I felt like Joyce wanted to go, but she said she wasn't ready to go, tho we did tell her we hated for her to go. But I told her to do what she wanted to. She said she wouldn't be gone long.'

"Jack, we started to buy a place last week and didn't hardly know what to do. We would have had to used about $1500. fifteen hundred worth of your bonds, but Dad thought he could have made a lot of it back. The place is about ¾ mile out west of M.W. on the hiway. 8 acres a 5 room house a sleeping porch & shower bath with a new heater in it & an apartment garage we could rent out & Dad said he could sell 3 blocks for a good price. It has gas & lights. It is $4000. So he went out there yest. & the Lady wanted to stay there 90 days and he said he thought she wanted to back out. We could borrow $2500 on it if we paid the rest. So we was going to take our bonds, but I told Dady I didn't want to use yours, but he said he didn't think you would care. Tho I guess that trade is off. Everbody thinks it best to buy now than wait. I was talking to Mr. Doss [banker] & he said he thought it best to buy now, as he thought it would still be high after the war, but I don't know what would be best. I know everthing is high now. I don't want you to have to bother about things like that now. I would like to be closer to town so Dad & I could work. I had rather work at something to pass the time. & to make some money & try to help save some of this cheap money.'

"Well I guess you are tired of that kind of stuff, but I don't have much news.'

"I sure hope you boys got your boxes for Xmas. I want to send another one soon as I can.'

"We hear the Japs are clearing out of Burma. I'll be glad when they are cleared out everwhere, tho I would be willing to give them their part of the world & stop this trouble. I can't see it will do any good no way.'

"The pecan crop is about over. We went yest. eve. a little while. The weather was too bad last week to do anything. I want to quit it & do something else.'

"I sent Curtis a V mail yest. If I knew you would get it sooner I would send them all that way.'

"We didn't hear from Loyd last week but I guess we will soon. We had one from Herchel written Xmas eve.'

"Paul Wesley is home on furlough. We haven't seen him. He's been convoying said he had made 12 trips across.'

"Well I'd better close & fix supper. So be careful and write soon as possible.'

For fear you didn't get my letters, we got your money orders. Dad bought you $250 worth . Two one hundred & 2 twenty fives. [U.S. Savings Bonds] You have got 26 hundred & 75 dollars worth in the bank now & we haven't got any of the ones you signed up for, but they said they were so far behind it might be a good while yet.'

"We got the box you mailed for Curtis but haven't got the one you mailed.'

"Well by & be sweet. We love you bushels.

Mother & all.

Jack wrote again on January 13, apparently from Mong Wi.

"Well here we are again on our fannies. Three days rest with 10 in 1 rations & pleanty warm Burma sunshine & cool almost frosty nights is just like a furlough.'

"Curtis & I just got thru digging us a two man foxhole that we know dang well we don't need & put our Texas flag on a bamboo pole about 40 ft. high. We want these liberated heathens to know that this is a Texas outfit.'

"Well I'd sure like to be able to tell you all some of the things we have been doing recently but I guess it'll have to wait. I can say that we are having lots of fun.'

"What is Jew doing now? I wonder if he is still working at W'ford. What do you all think about his chances of getting into the army?'

"We just heard our first news in about 2 weeks & they say that our people are back on Luzon. Well that's good. I'll be glad when they hit China. We should have them going right by then. Maybe it won't be long now. I guess that's a rough deal in Europe now. When have they heard from Rowe. Do you ever hear from Herchel & Loyd? I haven't for a long time."

"Well I'll write again soon. Everything & everybody is doing fine. All the M.W. boys are going good. I'm seeing to that.'

Write often

Love Jack"

Rowe Howard was a master sergeant in the combat engineers fighting the Germans in and around Belgium at this time. It was at about this time that he had his closest brush with death or injury. He told me in June, 2001 that he and three other soldiers were in a jeep traveling to a unit to inspect when they heard a German artillery round incoming. They stopped the jeep and jumped out. He went to the right with one of the men. The other two went to the left. The artillery shell landed between the men on the left and killed them both. Rowe wasn't wounded.

Mamma wrote on January 15,

"We went to Howards yest. They were doing pretty well. Mrs. H had asthma, but was up & going. They haven't heard from Rowe lately but I guess the mail can't come through very well now. We aren't getting much.'

"I don't have much news to write as there's nothing happening. We do have a lot of music now since we got the piano or noise rather. This bunch that lives across the hiway comes up lots. They have a guitaire & mandeline. Marvin came Fri. night. Tho I'm always thinking of you boys, whatever happens.'

"We will celebrate when you boys get home. We might dance all night – praising the Lord for he is the one deserves the praise.' [Mamma was always careful to guard against too much 'worldly' pleasure.]

"We are having some pretty clear weather now, since Joyce left. The weather was bad most of the time while she & Glenda were here. Dad is going to the camp this AM. to see about a job. Theres not many pecans left.'

"Well this is later at night. We had a letter from you, Curtis & Loyd today. Yours & Curtis were written the 12 & 14 of Dec. Was the letter you ask me to send you some stuff. I'm sorry I didn't get it sooner. I will mail it to you soon as possible & hope you get it ok. & soon.'

"Loyd said he had been pretty busy was the reason he hadn't wrote. I think Dad is going to work tomorrow as carpenter in the camp.'

"This is ten oclock. We are listening at the news it all sounds good. Better than it has in a long time.'

"Have you all been on another program. There was a captin said he heard some interviewed last night from the front in Burma. We missed that program last night, so I sure hated we didn't get to hear it. I bet we don't miss it from now on if we can find it at the right time.'

"We went down to Marvins tonight a little while. They played some music & wanted us to go. I guess you all will think we are having a big time by having so much music, but we just go cause the kids want to go & of course we like music too.'

"Well I don't have so much news to write, so will close & write some of the rest. I had a letter from Joyce today they were fine."

All our love & best wishes,

Mother & all

P.S. I hope you got you a deer. There sure are lots of wild animals over there. I told Marvin what you said about those little deer & his eyes bugged out. He goes deer hunting nearly ever year. I think down south with some one.

I wrote one of my rare letters at the end of Mamma's,

"Dear Jack,

"I have not much to say but I'll try to write something.'

"We made our pictures yesterday at school. I mean I did. Jr. and June was sick and never got to take their pictures. I'll send you and the other soldiers & sailor one too.'

"We are getting ready to go to school. I've been making good grades all year. I hope [I] do as good on Mid-term test as I do in six weeks test. We are taking our mid-term test this week.'

"We are learning to dance at school. We've learned about three steeps I think.'

Well Bye,

Bill."

This envelope was loaded. June and Daddy also wrote,

"Dear Jack,

What are you doing. I am fine and hope you are to, tho I was sick and had to stay at home. I wish Joyce would come up here.'

"I have found a place to take lessons, with Evelen.

Love June"

[Evelyn was Marvin McCracken, the fiddler's, daughter.]

"Tues morning, Jan. 16

"Dear Jack. 'how' you doing. We are all going strong.'

"I am going to Camp Wolters today to see about a job. I had rather work near home for less than go so far. I was just listening to the news it sounds pretty good now. You boys in Burma are getting in the news more every day now. It could be over this year 'I hope'. Jack we haven't rec. your box yet or any bonds. But they will come along after while. Old Mike came by to see us last week. He got 30 days. Goes back to Ft. Bliss the 29th. He is tank gunner, a corporal, 5 overseas stripes on his sleeve.' [Mike was Uncle Bill Holder's oldest son.]

"Well Jack you & Murphy take care of your self and be good. Write as often as you can. I gotta go.'

"With all my love.

Dad"

During their bivouac within a day or two of these letters, Curtis decided the troop needed some fresh meat. He combined his experience as a cook, his rifle expertise and trading skills learned at his Daddy's side to secure fresh meat for the table. With permission from Jack and with Billy Ralph Aaron, a Kachin as interpreter and scout, two mules and several parachutes, he set out to locate a Burmese settlement nearby. Through the Kachin, Curtis started his trade talk.

"Chief, will you trade cattle for these parachutes?"

"Maybe."

Curtis countered, "I will give all these parachutes for four yearlings."

With a frown, the native trader answered, "No. Two."

This wasn't going to be easy. Knowing that it would be difficult to lead four animals to camp, Curtis countered,

"O.K. I will take three beeves for the cloth. This is my last offer."

Satisfied that he had gotten the silk for half what he was prepared to give, the chief answered, "I give three for cloth. You win."

Curtis won because the Kachin had whispered a word of advice into the chief's ear. The reputation of the Kachins from Northern Burma was widespread and hadn't escaped these remote Burmese villagers. Curtis didn't understand what was said, but it convinced the chief that he would be well advised to overcome his reluctance and allow the beeves to be slaughtered for food.

Their plan was to catch the cattle and lead them back to their overnight camp to slaughter. However, the wild cattle were spooked by the strange people and even stranger looking pack mules. They couldn't be caught.

Curtis took the pragmatic approach, shot the yearlings, field dressed them and with the mules packed them to camp. It was good he didn't get four beeves. They would have had to leave one of them there. It was late in the day and they had to employ torches made from parachutes to light their way back to camp.

The next day the chief arrived to protest a violation of their agreement. He tried to convince squadron officers that he intended to trade only two of his cattle herd and deserved more in payment. However, the testimony of the Kachin Ranger revealed that the chief was merely having a go at increasing his largesse through deception. Also, with some political acumen, Curtis had seen to it that the squadron staff officers had shared their roasted bovine bounty. The chief was sent back to his village with thoughts of his own inadequacies and a lesson in bartering skills.

The fresh beef, barbecued over an open fire was a welcome relief from the C and K and 10 in 1 rations that were a staple for American fighting men.

Dealing with the terrain was enough of a challenge without the added burden of dealing with hostile natives. The Americans also needed the good will of these people if they expected to get help with the Japs. Traveling thereafter became so treacherous that the effort at scavenging or trading for food was abandoned.

Jack wrote from the bivouac area again on January 19,

"Dear Folks

"How are you all? We are doing fine. We have had one of the easiest weeks I have had since I came into the army. Just dozing around in this warm Burma sun during the day eating our uncle's best 10 in 1 rations & its just cool enough at night to be comfortable . We saw a little frost one morn. Up in the mts. We use parachutes for sheets & sleep like kings. We have swaped for 3 calves; about 200 lbs each, & the troop has had some real steak. We have built bamboo shacks all over the place & it looks like a Gypsy joint. Everybody is getting fat again.'

"I got a letter from Loyd the other day. He seems to be running along about as usual. He kinda talked like they might be about ready to move out. They may go to the Philippines. Maybe we will meet them down in French Indo China or someplace. I think Toughy is down here off the China Coast. They say they are giving Formosa a going over. Maybe it won't be long now.'

"They have a radio here at Sq. HQ that we keep going all time & at night I just lay in my bunk & listen to music & the news over our sound power telephone. This is really the life.'

"Well I gotta write my howney Sanchez. She might get mad weeth me.'

"You kids write once in while. The letters I write to you all are to all of you. Jew what are you doing?'

"Write soon. Love, John"

Mamma wrote three days later,

"We had a letter from Loyd and Herchel this week. Herchel was still at the same place. Said he might get to see Loyd. He had been with Tom Vance. I bet they had some time. Was the first time Tom had seen anyone he knew since he went across.'

"We got a notice for Vercia to report at Austin the 24th so Uncle Bill came down & he & I took her to W'ford to get things fixed up & get her some more cloths. It is a school and they said it sure is a nice place & the way they learn is wonderful. I sure hope it does help her & I think it will.'

"The war news sure sound good now. The russions advancing 40 mi. a day & the Japs getting theirs too. Thank goodness.' [Mamma's pacificism had mellowed considerably with her boys in harm's way.]

"Well Jack honey, for fear you might not have got my letters, we got all the money you sent $270 & $40 for Xmas. Aunt Mattie heard from Norris yest. Said he had been too busy to write, fighting the Germans.'

"June is going to get to take music lessons at school. They are going to start giving piano lessons down there.'

"R.C. is still working but may not be long. We can't tell. He wrote us a card & said he was going to have to be examined. They are going to reclassify about 100 in Parker Co.'

"I am having my teeth filled. I've been two trips & have to go again. It sure is expensive. Will cost $35. Dad kept $11.00 of your money to pay on them that was left after buying your bonds.'

"We hear from Joyce regular. They are ok.'

"We got a notice from the government that you & Curtis can't write right now. We sure hate it but please don't worry for we think we understand. They said they would let us hear twice per month.'

"Well dear, I have to close for now. Tell Curtis I'll write him tomorrow.'

"So be real careful. We love you.

Mother & all"

R. C. had given up on school and had begun working at a dairy near Weatherford. He expected to be drafted, but his work might delay his induction.

The Second Squadron hit the trail again. They knew the Japanese were just a few days away. Fighting had been underway for several days. The Chinese had joined the 475th at Namhpakka. The area was infested with well-supplied and dug-in Japanese. The infantry regiment had already entered the fight. Jack was about to get his long-awaited wish. (Marsmen in Burma)

The last leg of the march began with men well rested, well fed and well supplied. They were close. The mountains were much higher and would require more effort than the ones behind them. Many of the men had become addicted to cigarettes. Their use of this narcotic was encouraged by official policy. The men were provided the cigarettes free. I wonder how much the tobacco industry has profited from this policy, not only in immediate profits, but in life-long addiction. Their habit exceeded the limit for maintenance of good conditioning. These men would have the most difficulty with the remaining segment of the march.

An entire day of marching would be up hill. The effort became so exhausting that, to keep the serials together, a schedule of ten and ten had to be adopted. Marching only half the time made for very slow progress. I was told by Charles Baker, then a platoon sergeant, fifty-seven years later, that his whole life focus centered on being able to take that next step and he wasn't a smoker. Curtis told me that he never experienced this kind of agony. He and Jack never smoked and always kept in excellent physical condition. Had any American fighting unit been required to endure the same conditions on their way to combat? I'm not aware of it. Even with this torturous endeavor, they knew the enemy was only a few days away and they pressed on.

At this point I must insert the story about the Second Squadron surgeon's prowess in pack toting. Dr. Jay Huey was a young man just out of medical school when he joined the 124th Cavalry at Ft. Ringgold. He was also one of the smallest men in stature according to testimony of men from Troop F at their reunion in April, 2002. Dr. Huey carried the largest pack in the regiment. They said it reached from his knees to above his head. He made the entire march this way.

I had talked to him at the reunion two years before. I knew him and felt comfortable calling him at home to check this out. I asked him what he was carrying in that pack. He told me that he, like the others, had discarded all unnecessary supplies and equipment. His only eating utensil was a spoon. He was satisfied to eat out of ration cans. His extra load was his medical kit to which he had to have access as he moved from one troop to the next, treating these warriors' nicks, bruises and illnesses. I didn't ask, but I would wager that Dr. Huey was no smoker, just a dedicated professional with irrepressible will and courage.

The mission of the Mars Task Force was now officially described in orders to its various units. The Japanese had established a base of operations on the Burma Road near the town of Namphpakka some twenty miles to the east of Mong Wi. Each combat team, consisting of either a battalion of the 475th or a squadron of the 124th were assigned a hilltop overlooking a large valley filled with Japanese defensive positions atop of hills within the valley. The Japanese supply dumps were scattered throughout their positions. To the north, east and south were escape routes. The mountains to the west had been their idea of protection. They were surprised to discover allied troops in large numbers on these mountains and in their propaganda, attributed their arrival on the scene a result of parachuting. The Marsmen were to destroy Japanese installations, disrupt communications and block their escape routes.

After much debate over which squadron would lead the remaining days of the procession, the Third Squadron was chosen. The next day the First Squadron would follow. Troop F, along with E and G Troops of the Second Squadron was to stay behind with the Brigade forward echelon as guard and reserve. They would be called into action only when needed. (Marsmen in Burma)

As they prepared to form their march column, Jack complained to Curtis,

"Well, here we are, ready to kill some Japs and guess what's happened. Second Squadron has been assigned as reserve. It just looks like the more you want something, the more things get in the way. You just wait and see. By the time we get into it, the Japs will be run off or killed and the most we'll get to do is shoot a few runty stragglers. We got the hind tit again."

All Curtis felt like doing was grunt in agreement. He had heard similar complaints for three years.

Reference has been made to the respect that Jack's men had for him. A story told by Alvin Moberly, then a Pfc., illustrates the kind of leader that he was. Pfc. Moberly had been assigned the BAR as his combat weapon. It weighs four times as much as the carbine, Jack's weapon. He told me that Jack saw him struggling up one of the sheer slopes on this last leg of the march. He traded weapons with him for the remainder of the day to give Pfc. Moberly some much-needed relief. According to Mr. Moberly, this was just an example of the things he did for his men.

Mamma and Daddy wrote to Jack on January 21,

"We just got back from seeing R.C. He was ok. He's just been home one time since he went to work. I guess he will be home next Sat. night to go with us to Uncle Bill's Sun. Mike leaves next Sun. night.'

"Well I didn't think I would tell about me going to work. I'm working at the classification building in the Camp. Dad is doing carpenter work. The kids & all of us leave home at the same time & Bill & Jr. get home a little while before we do. June stays with Aunt Rosa till we come by from work. I enjoy working. It is easy sit down a lot of the time. Sallie Howard , Lizzie Massey, Bernice Cowley & Mrs. Sims works with me & a few more women. We have a lot of fun. Oh yes Mrs. Hobson works. She's the one we have fun with. Lizzie teases her all the time. They all pop off all the time. Some time Lizzie & Sallie fusses, then forget it in five minutes. I get 50c per hr. Dad gets 75c.'

"I met Mrs. Wilson the other day. She works in the sewing & mending building & we use the same rest room. She was so glad to meet me & I like her too. She said Little Red wrote her if she ever saw me to coment me on having such nice boys. Said he was back in Troop F, thanks to Lt. & Sgt. Knight. She interduced me to several women like I was somebody extra & was bragging on you boys & of course Lizzie Massie & some more was there & heard it & that made me feel kinda extra & of course it made me happy to have some one to brag on you boys more than anything else, for I know you are the best in the world.'

"This war news sure sound good & it's ever where too. As yet we haven't heard any news about the Mars Task Force. Mabe they are on a secret mission. I'm not saying anything about them for I don't know anything to say.'

"We had some sunshine today for the first time in a week or more. It rained most of this week and now looks like we still have some cold weather. It's a cold norther. I hope it isn't cold where you & Curtis are, for I know it would be hard on you or too hot either.'

"We hear from Loyd & Herchel quiet often. Herchel said he might get to see Loyd & he was with Tom Vance one day. I bet they had some fun.'

"Aunt Mattie heard from Norris. He was ok. Just been trying to take care of himself. I think he's in Luxingburg or was. Fred's in France & Johnnie I think left from San Francisco, Calif.'[The Watkins brothers father, Paul, had fought in France during World War I.]

"Tell Curtis we got a letter from him to Joyce & as usual Jr. opened it thinking it was to us, so I just had to read it. Was written the 30th of Jan. [She must have meant December.] I sure was glad to hear from him.'

"We are sending your box tomorrow. Just couldn't get the stuff together any sooner & can't send all we wanted to. I haven't baked you a cake yet. I'm a little afraid a cake would ruin. Write & tell me if your cake was ruined I sent so I will know. That is when you can write. I might risk it anyway. I will send Curtis a box too. I hope you get them.'

"I will close & let Dad write some.'

"We love you bushels. Our love & best wishes.

Mother & all."

"Hello Jack,

"Will say a word. Hauled wood today. Am working at Camp Wolters at the carpenters shop. Will make $39.00 per week. I like it pretty good. We got the notice from the government about your mail. We may not hear from you directly for some time. But I hope it wont be long'

"The war news sounds good now. Them Roosians are gone berserk.'

"I hope the Nips pull out of your C.B.I. They had better before it is to late. Tell Murphy I said hello and good luck to you both. Take care of yourselves and be careful. With love

Dad."

After their rigorous march, Second Squadron arrived on January 28 at their assigned hilltop and dug in. They had a panoramic view of the valley below and could observe military action in all directions. The battle zone was an area approximately five miles long setting astraddle the Burma Road and roughly three miles wide. They had walked approximately 300 miles, with a week of rest at Mong Wi, since December 19, an unprecedented feat for a military force of this size in this type of terrain. (NACP, R.G. 407)

Prior to the arrival of the Second Squadron, the Third Squadron initially and then the First Squadron had joined the battle with the 475th Infantry. Excerpts from the 124th Cavalry S-2 reports indicate some of their combat encounters,

January 21:

"3d Sqdrn continued attack on objective ... and secured entire hill with the exception of southeastern nose. Sqdrn dug in ... and repulsed two JAP counter attacks which were supported by heavy JAP arty fire. 1st Sqdrn arrived with 613 FA which immediately went into position...opening up on a 19 man JAP patrol, then on a truck conv. on the Burma road in vicinity of NAMPAKKA. (results unknown.) "A" & B Troops were immediately dispatched to reinforce 3d Sqdrn's and went into position.... 2d Sqdrn enroute from Mong Wi.

Casualties – 2 killed 5 wounded"

January 22:

"B Troop, supported by 613 FA Bn. attacked and cleared southeastern nose of ridge at SO 698468. Fifty dead JAPS have been buried in and around 3d Sqdrn. positions to date. 613 FA fired on estimated 300 JAPS in vicinity of NAMPAKKA registering directly in on JAPS – results unknown. "A" Troop reverted back to control of 1st Sqdrn. and went into position as shown on overlay. Elements of 114th Regt contacted Regtl C.P. arriving from N.W. 2d Sqdrn enroute from MONG WI.

3 casualties – 1 killed

475th Infantry S-2 Report, January 22:

"First Battalion patrols...made no contact. Scattered bursts of machine gun fire from the area of Man Sak fell on "B" Company positions of the south end of the paddie at daylight.'

"Second Battalion patrols operating south of the perimeter toward the village of Loi-Kang were hit by Japs from dug in positions along the ridge. Patrols to the east reached the Burma Road without contact, and laid mine fields on the road as shown on overlay.'

"Third Battalion patrols (I & R Platoon) reconnoitered east to the road and encountered estimated two Jap squads which withdrew after being fired on by machine guns and mortars. The patrol reported the crater blown in the Road on the 19th had been repaired. The Battalion perimeter was extended to permit observation of the draw to the north and east of the hill.'

Morale – High'

Status of evacuation: By native litter bearers to strip, and by plane from strip to rear."

475th S-2 Report, January 23:

"First Battalion patrols ... made no contact.

Second Battalion continued to meet resistance from Jap dug in positions on north end of Loi-Kang Ridge. Patrol...contacted Jap outpost. Patrol to road set additional mines, and reported two Jap trucks destroyed by mines previously laid.'

"Third Battalion patrols to north and east made no contact.'

"Morale – High'

"Intermittent artillery fire continued to fall on the drop field thro[ugh]out the period, and a few scattered rounds hit the Regimental CP.'

"Status of evacuation – No change"

475th S-2 Report, January 24:

"First Battalion patrols to the south and west...made no contact.'

"Second Battalion probed Jap positions on north end of Loi-Kang ridge, and continued to meet determined resistance. An enemy Tankette was destroyed by mines previously laid on the road. Additional mines were laid, and an effective Road Block was constructed by using the wrecked Tankette, and trucks destroyed on the 22nd.'

"Third Battalion patrols located an abandoned Jap Aid Station and ammunition dump... A supply of 105mm and 150mm shells were found in the ammunition dump.'

"Morale – High'

"Contact established with 114 Chinese Regiment and plans coordinated."

(NACP, R.G. 407)

Elements of the 475th were having a difficult time with Loi-Kang ridge. It would remain a matter of concern until February 2.

Excerpt from 475th S-2 Report, January 29:

"Second Battalion listening post of a reinforced platoon...heard one tank attempting to move north on road during the night. This target was fired upon by the platoon, and two tanks, which were observed on the Road adjacent to third Battalion positions moving south, withdrew to the north when they saw fire being placed on the first tank. Night ambushes on the north and south ends of battalion perimeter each killed 1 Jap. A platoon of Kachins established an ambush...at 1030 and remained in position until 1330 without contact when they returned to the Battalion area. P and D platoon patrolled to the Road and blew a large crater in center of road between tankettes and trucks previously destroyed. They again found that the Japs had disarmed mines and booby traps on the road. Said mines and traps were rearmed. No enemy artillery fire fell on Battalion positions during the period.'

"Third Battalion fired on small group of Japs infiltrating westward in draw between their position and those of 124 Cavalry during the night. It is believed these Japs are attempting to reach the drop field area in order to steal rations. No enemy contact during period.'

"The 612 FA Battalion received 12 rounds enemy artillery fire believed to be 115mm and 150mm at 1500 hours. Two American casualties."

Excerpt from 5332d Brigade (Prov) G-3 Report, January 29:

"1st Sq: A Tr, with support from B and C Trs, successfully attacked Jap platoon..., killing 34 Japs and occupying this ground. B Tr extended Sq perimeter.... C Tr unchanged.'

"2d Sq: Moved from HO-PONG...and were released from brigade Reserve to CT 124.'

... "CT 124: Continues consolidation and occupation of positions.

Effectiveness of co-ordination between CT 475 and CT 124 tested and proved satisfactory."

Jack and Curtis were getting closer to combat status with the Second Squadron removed from guarding headquarters.

124 Cavalry S-2 summary, January 31:

"1st and 2d sqdrns continued to consolidate positions and counted 35 dead JAPS as result of previous nights Banzai charge. Patrol hit JAP platoon.... Two men missing from this patrol. Arty fire knocked out one tank and hit a JAP gun position.'

"Casualties – 2 men missing"

The Second Squadron had arrived into this very active war zone. F Troopers were able to observe their first combat, but had remained in reserve. Quoting from "Marsmen in Burma,"

"It was on the 28th (January) that the 2nd Squadron arrived and got their first baptism of fire as Jap artillery, probing for our own, dropped shells around them as they came over the hill west of the valley. F Troop set up between 1st and 3rd Squadrons. E and G Troops were close behind. A job was in store for them, but they had to sweat out the 'go' signal from higher headquarters.'

"The 28th was a big day for the China and India-Burma Theaters. The first overland convoy to China in three years passed within a mile of front lines up the Burma Road north of us to head into China. Secrecy had surrounded its formation at Ledo and its progress over the Ledo Road to the now cleared and open junction with the Burma Road. Timing was perfect. It had been poised ready to pass north of the retreating foe as soon as it could. The Japanese stranglehold on this famed lifeline had been loosened so that not only supplies themselves could go by road, but a stream of trucks could pour into China to help with its internal transport problem. The world's longest and most fabulous pipeline could now push into China to provide much needed fuel for planes and vehicles.'

"The most substantial part of the supplies to go into China came from the United States. The little closing link of the longest supply line in the world had been fastened into place. We were helping to weld it securely. The supply line extended nearly 14,000 miles – a distance greater than halfway around the globe at the equator. Cargo, coming by sea from the United States, arrived at Calcutta. Loaded there onto freight cars, it had to be reloaded again when shifted to narrow-gauge [rail]road halfway enroute to Ledo at the end of the line in northeast Assam, India's eastern-most province. Here the Ledo Road and truck portage began and nearby was Dinjan, from which supplies were flown to bases in Burma."

Controlling the Burma Road to Rangoon and further south would shorten this supply line by hundreds of miles and save valuable time. Fighting in these mountains was an extremely important part of the plans to oust the Japanese from Burma and defeat them in China. After the exhaustive march, the 124th joined the 475th on the Burma Road to accomplish the goal.

General Wiley, commander of the Mars Task Force, called it, "...the most hazardous terrain in Burma ever traversed by an American fighting unit." and "... a magnificent job of marching." As a member of the famous Merrill's Marauders, he had the credentials to back up his statement. (Marsmen in Burma)

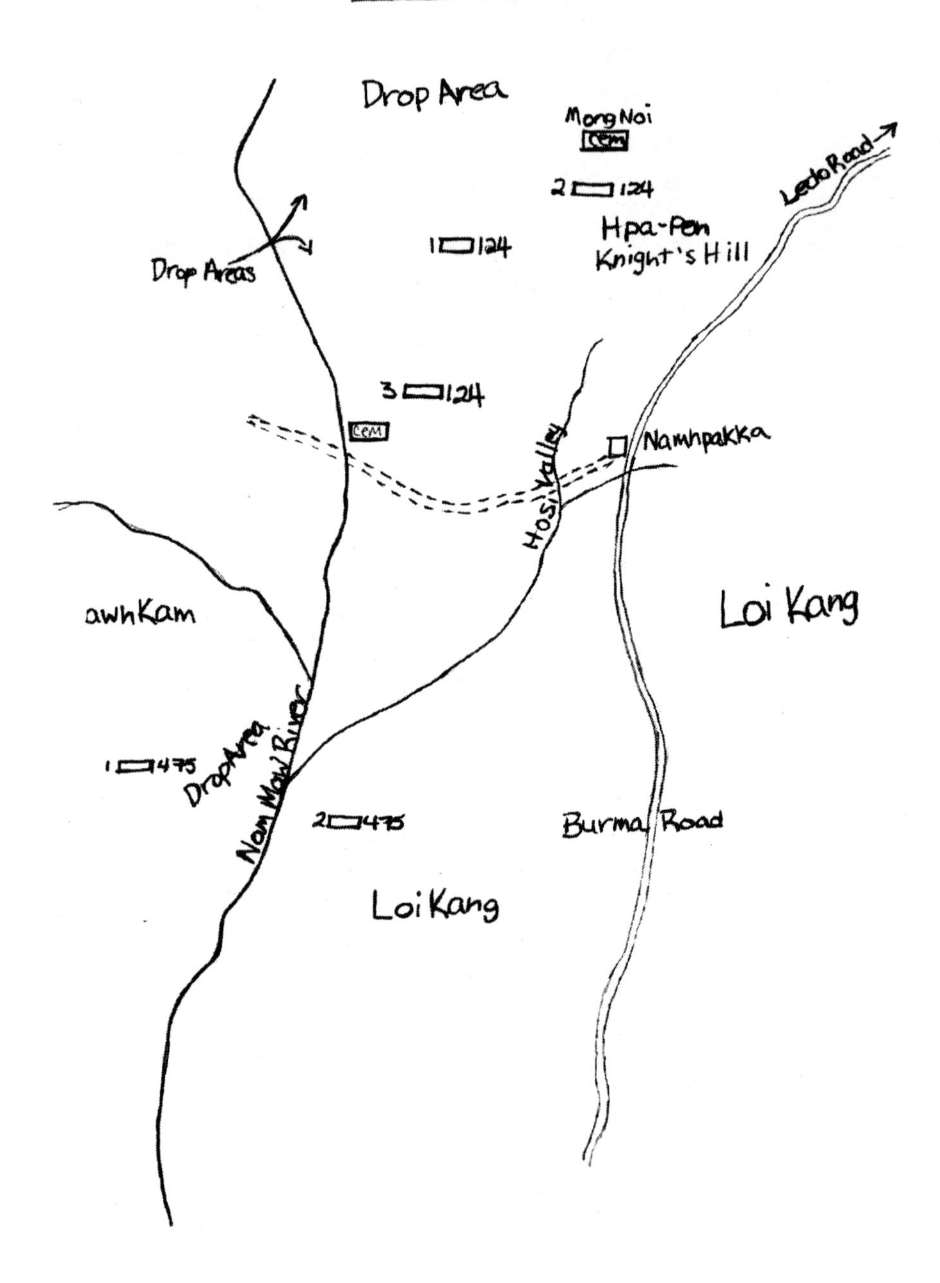
Mars Task Force Battle Area
Drop Area
Mong Noi
CEM
2 124
Ledo Road
Hpa-Pen
Knight's Hill
1 124
Drop Areas
3 124
CEM
Namhpakka
Hosi Valley
Loi Kang
awhKam
Nam Maw River
Drop Area
1 475
2 475
Burma Road
Loi Kang

CHAPTER 20

BEFORE THEIR FIRST AND LAST BATTLE

SECOND SQUADRON WAS positioned between the First and Third Squadrons to guard Regimental Headquarters and to serve as a reserve unit. Fighting was heavy all around them day and night as a combination of British and Chinese soldiers joined the men of the Mars Task Force to take the territory now infested by a large Japanese force sitting doggedly astride the Burma Road. This was also intended strategically to prevent Japanese divisions to the north in Burma and east in China from using the Burma Road as an escape route from massive attacks by the British and Chinese. (Marsmen in Burma)

Mamma was still pumping out the letters to her boys. She wrote Jack on January 25,

"How are my darlings. Hope you are well and you & Curtis both are well & doing fine. We are all well.'

"We had some news yest. from over your way. The heading was Americans Cut the old Burma road and about some specialy trained troops traveling 300 mi. through the jungles before going into action. So we thought it might have been you boys, if so I know you had a tough time but feel like you are both ok. Looks like the Japs had as well give up. They sure are catching it on the Philippins now.'

"The news all sound favorable.'

"We had a letter from Loyd & Herchel yest & today. They were ok. Loyd said he hadn't heard from you boys lately. I hope it won't be long now till we can hear from you & Curtis.'

"I mailed you & Curtis a little box but couldn't send much as it couldn't weigh over 5 lbs. Sure hope you get them. I'm afraid you all didn't get all your boxes I sent – three or four since October. Sure hope you did get them. I sent some boxes of pecans.'

"Anything you want be sure & tell us. We didn't send Curtis as many razor blades as we did you, cause Joyce had sent him a lot when she was home.'

"We finialy heard from your box. It's in Ft. Mason, Calif. We had to fill in a blank with our name & mail back before they would send it.'

"We are going fine at our job. Tell little Red Wilson I see his Mom ever day. She is ok.'

"Vercia went to Austin yest. Aunt Mattie & Jr. went with her. Jr. said it was a nice place. She was anxious to go.'

"Well June is wanting me to go to bed with her. So by for this time & be sweet & real careful.'

"All our Love & prayers are with you all.'

Mother & all."

June's note that accompanied Mamma's letter,

"Dear Jack

What are you doing. I am fine and hope you are to. I made 70 on arimetic. I can't think of anything to write hardly. At school we are going to have a Rythmn band. My teacher's name is Norma Harris. I have 3 teachers one of them is Norma Harris and Mrs. McDonald and Mrs. Crounse."

Love June"

Mamma wrote again January 30,

""We rec'd your letter today- of the 13th. Sure glad to hear from you & you were all ok and hope you are still the same.'

"We had a card from the Weatherford Democrat today wanting to know something about you boys & said they had an action picture of you that we could get if we didn't have one. So Jr. is going tomorrow & take a letter Dad wrote about you boys & to get the picture. We sure are anxious to see it.'

"R.C. was home Sat night & sun. He's still working at the dairy.'

"I'm still working but don't think I'll work long.'

"Oh yes, you ask about R.C. His boss thinks he can keep him. I sure hope so. The man fired his own boy. He got married & got to where he wouldn't get to the barn till they would get almost done milking. So his Dad said he was going to send him to the army.'

"Yes Howards heard from Rowe. He is in Belgum I think.'

"Aunt Mattie has heard from Norris too and we hear from Herchel 2 & 3 times a week now & get it in about 5 & 6 days. We heard from Loyd last week & he was ok.'

"I see little Reds mother ever day I work. I'm glad I can tell her what you said about the M.W. boys being ok.'

"I don't have anything new to write. Everbody is working or going to school.'

"I'll be glad when this is all over and everbody get home & settled down to natural way of living.'

"Glen Day was wounded in the leg & come to Calif. So Claudene & baby went out there & they were looking for them home yest.'

"Well, I'll go to bed so be real careful & write ever chance.'

All our love,'

Mother & all."

Daddy wrote the next day, Wednesday morning, Jan. 31, '45,

"Dear Jack & Curtis,

Will drop a line. We are doing all right. Just heard on the news about the Mars Task Force inflicting heavy causlties on the fleeing japs. I hope you boys will be careful. The news sounds good now. Buzz is taking another boat ride.'

"Tuffy is at P.H. He saw Tom Vance. Got a kick out of it. I am working at Camp Wolters carpenter shop. I like it pretty good.'

"Bessie got word from Son that he lost his leg between the knee and hip. He got hit in the Battle of Leyte. He will be in the States pretty soon. Earl Gilbert & Alex McQuary are missing in action. They were in Belgium. [Rowe] never got caught. I guess he was to fast on foot. Connie Marie is working at CW. Will have to go work, so good luck with lots of love.'

Dad"

'Son' Finger was our first cousin, the son of Bessie, Daddy's sister. He was in a glider unit. After they landed, they were fighting their way through the jungles and a Japanese sniper, firing from his nest in a tree, hit him just above the knee. He came home and never let his handicap slow him down. One of my most vivid memories of the time was watching him vault over the garden fence with his crutches. He would come by the house purposely to get Mamma to fix a freezer of ice cream. It was usually peach or banana. The government gave all amputees a new car. 'Son' apparently thought it had only one speed, wide-open. The few times that I rode with him, I learned to deal with fear.

Jack was still thwarted in his obsession to get at the Japs. His impatience had diminished because he knew that it could be only a matter of days or even hours before Troop F would be called upon. The wheels began to turn anew on January 31. The Mars Brigade was stretched thin and a decision had been made to throw the Second Squadron into the attack. Men of Troop F recognized the dangers of Japanese artillery and prepared defenses as best they could. While cutting tree limbs to fortify his and Jack's pillbox, Curtis made a misstep and fell from a tree, scraping his rib cage. While getting first aid from 'Windy' Andrews, a photograph was taken that was included in the photograph section of Marsmen in Burma.

The remainder of the story will be told through the use of material gleaned from "Marsmen in Burma" by John Randolph, from archive records of the 5332nd Brigade and interviews of F Troopers. For the first two weeks, since January 17, the 475th had done the lion's share of the fighting, as all of the 124th converged on the hills west of the Jap positions. The story will shift to the 124th, even though the 475th continued to contribute equally to the fighting.

On January 29th, Troop A of the [First] Squadron with one platoon from C Troop assaulted a Japanese infested hill. They were supported by B Troop and by artillery fire. They attacked at 1100 hours. John Randolph tells the story,

"At the prearranged signal of a burst of machine-gun fire, the attack started. Terrain was fairly open. There was some knee-high grass and small trees. From their line of departure they came up as one man to run 200 feet into a draw and climb 250 feet, almost straight up in some places, to the crest of the Jap-held hill.'

"Firing as they went and still on the run, they faced dugouts all over the crest of the hill. Each of the troop's platoons had a two-man bazooka team attached—smooth, cool operators who blasted the well-protected emplacements of Japs who would not come out and fight. Organized grenade attacks took care of others, as did small-arms fire. Few Japs stood to fight and some were mowed down as they tried tardily to escape.'

"A Jap reserve unit attempted to pull around and flank their attack. Locke called for help, and B Troop moved in to block this attempt and let the show proceed according to plan.'

"In less than an hour of fast but meticulously managed fighting, the hill belonged to A. Gains were consolidated and the men were deployed in defensive positions. At 1155 Locke reported to Colonel Ripstra that the hill was his.'

"They counted 32 bunkers on the hilltop and 34 dead Japs. Troop A's casualties: two killed and seven wounded. One of those severely wounded was T/4 Frank Ferrante, of New York City, who had come to line duty a few days before after two years of kitchen duty.'

Records of the 124th Cavalry mentioned that one prisoner was taken and that continued attacks of the Japanese during the day brought a total of 124th casualties to eight killed and twenty-one wounded.

The Second Squadron moved from Ho-Pong and were released from Brigade Reserve to CT 124, their regiment . The G-3 report for the 5332d Brigade on January 30 stated that the 475th and the 124th continued to consolidate and occupy their positions. Consolidation of the two units had been tested and proved satisfactory. These reports were prepared by Ralph E. Baird, Lt. Col. Cavalry.

January 30th brought a great deal of Jap attention to the position of Third Squadron CP. The CP was located just under the crest of a hill on the western slope. The Japanese fired 90 mm mortar and 105 mm artillery into the CP area. One man was killed with eleven wounded. Squadron Commander Colonel Hazeltine quickly relocated his CP. Jap artillery also destroyed an ammunition dump near the I Troop position.

On this day C Troop was to have its share of bad luck. It sent out a platoon on reconnaissance. John Randolph describes what happened,

"A reinforced platoon from C Troop, on reconnaissance, hit Jap opposition almost immediately after it left the perimeter. Lt. David H. Shepard was hit. Lt. Erving Koffer, from Brooklyn, knelt beside him as the patrol took defensive positions. Both of them insisted that the patrol go on. They were toward the rear, and Koffer gave assurance that he could take care of Shepard. When the patrol returned, neither of them was there. The next day they went out to find

Shepard's body, and near it was Koffer's helmet and field case. It was the only evidence that we were to have that the Japs may have taken a prisoner. If he was killed, his was the only body we were not to recover." (Marsmen in Burma)

According to the S-2 summary for the 124th for January 30, the First and Second Squadron continued to consolidate positions and counted 35 dead Japs as a result of the previous nights Banzai charge.

The Nam Maw River ran through the valley between the Mars Task Force positions on the west and the Japanese positions in the hills of Loi Kang Ridge that lay between the Marsmen and the Burma Road. Because of heavy jungle growth the men were able to find safe places to secure water. They filled their water bags from a clear running stream and packed them to the troops' positions on their trusty mule comrades. Some of the men found the time and opportunity to locate an area in the river below the water hole to take a bath and wash their clothes after more than a week of combat.

The perimeters of both the First and Second Squadron were counter attacked during the early hours of January 31st. Our artillery and mortars helped drive the Japs off after they had killed six and wounded nine of our men. Reconnaissance patrols mopped up Jap positions on high ground around Kong Song. The 613th Field Artillery hit one of two tanks during the day. They received air support with a bombing mission.... John Randolph recorded this sad tale.

"A direct hit of a Nip 150mm shell caught Capt. Maurice H. Hunter, of Eugene, Ore., CO of E Company, and his executive officer, Lt. William E. McQuirt, in their foxhole and killed them both."

The following report was prepared by Captain Phillip R. Neuhaus for Major Jordan, Commander of the Second Squadron. It is taken from archives, Mars Task Force and is summarized.

"Major Jordan began to position his troops and planning for a February 2 assault on Loi Kang Ridge. On January 31, he assigned one platoon from Troop E to go into position with the Chinese on the north perimeter of the battle zone. The squadron headquarters troop and the 81 mm mortars were moved to the village of Mong Noi.'

"On February 1, Major Jordan ordered Troop E to pull out of First Squadron position in the morning, one man at a time. Two platoons of Troop E were moved to its jump off position and would coordinate with their detached position.'

"The heavy machine gun section of Troop F pulled out at daylight with their guns camouflaged in mortars to join Troop E. Troop G moved to the assembly area at Mon Noi. Troop F remained in position until 0100, February 2, when it moved to Mong Noi. Guides led Troop F and then led all troops to the line of departure. A screen from the Intelligence and Reconnaissance Platoon shielded troops until the attack was launched.'

"On the morning of February 2, artillery and 81mm mortar units was ordered to fire at pre-arranged targets in the Japanese held zone. First Squadron was to assist with 60 mm mortars, machine guns and small arms. Artillery fire was then to be concentrated on the target as First Squadron makes a feint on the south of the enemy position. The Chinese and the Troop E platoon were to wait until First Squadron slackened or until the target of opportunity appeared, then was to fire on the ridge to keep the enemy pinned.'

"At H hour of 0600 the attacking force, spearheaded by Troop F left the line of departure. First Squadron 81 mm mortars was to begin to rake the draws on the face of Loi Kang ridge. Troop E, would be supported by the heavy machine gun section of Troop F [led by Sergeant Truman Owens] as Japanese positions became defined. Forward observers with the assault troops would direct fire on points of stiff resistance. [Lt. Leo Tynan was assigned to Troop F.] Additional mortars were in position to assist with orders by radio from the assault troops. Mortar and artillery fire from First Battalion, Eighty-Ninth Chinese was to be directed on the northeast face of the objective hill.'

"Major Jordan also assigned Kachin guides to the troops. They knew the exact locations of positions. His orders also gave directions on the use and feeding of mules, the amount of ammunition on hand, and the plan to evacuate casualties. His orders gave further instructions concerning the veterinary station, burials, salvage, captured material, prisoners of war and directions for service troops relative to rations and ammunition supplies."

Second Squadron prepared additional information and instructions contained in its Field Order #2. Excerpts from that order follow.

"Company strength of 1st Bn, 4th Regt, Japanese is estimated at 50 men for one company, and 80 men each for two other companies. Reenforcements may have increased their figure. The Squadron objective lies within this Battalion area. Enemy strength in the Hosi-Namhpakka area is estimated at 1500 men. It is believed that there are one or more medium Jap tanks in this area armed with 77mm guns. The enemy is well equipped with automatic weapons and knee mortars. All their personnel may be expected to be well dug in. The enemy mission in this area is to delay the cutting of routes of withdrawal of their forces from the North. The individual soldier's mission is to defend to the last man. He appears to be well fed and to have an abundance of ammunition."

Field Order # 2 from Major Jordan continued with orders for assault troops to be Troop F, with Troop G on the left. Each troop was to attack on a 300 yard front. Troop E, with attached troopers, was to support the initial attack. Two bazookas were attached to each rifle troop. Each unit was responsible for protecting its own flank. After the objective is seized, Troop F will reorganize and generally face west with Troop G facing generally east. Troops F and G must be prepared to hold the objective without Troop E.

Most of the day on February 1st was devoted to re-positioning units of the 124th and preparing for the assault on the Japs the next day. Japanese bombardment from heavy artillery and mortar continued through the day and increased after dark. Several men were killed, including Colonel Thrailkill of the 475th who was with his troops checking on a mortar OP. He had been wounded in the Tonkwa action.

The P and D section of the Second Squadron began construction of an observation post overlooking the hill that was to be the scene of attack. They were observed by the Japs and were shelled heavily. They had to re-locate the O.P., exercising more caution and concealment.

Patrols from all Squadrons of the 124th probed enemy positions in the vicinity of KAWNGSONG. The 613th Field Artillery Battalion scored a direct hit on a tank.(NACP, R.G. 407)

In late evening of February 1st, Second Squadron was positioned between the First and Third Squadron on the western perimeter of the battle zone. Their position was receiving part of the Japanese bombardment. Jack and Curtis were dug in immediately behind the perimeter of Troop F. A runner approached their foxhole and told Jack that one of his F Troopers had been hit by mortar. With caution from Curtis, Jack went to see about him. It had been a direct hit. The trooper was dead, the first combat casualty of Troop F.

At 1 o'clock in the morning, Second Squadron moved from its position to the west in a circuitous path west and behind the lines of First Squadron to its jump off point. The men moved cautiously in small groups and one or two at a time to confuse the Japanese. By 2 a.m. they had established a new perimeter to the north of the objective hill. (Marsmen in Burma)

Jack and Curtis had had little time for casual conversation during these intense days of preparation and dodging artillery and mortar fire. This was in major contrast to the preceding six weeks of arduous marching through tranquil and treacherous mountains. Each task had become more dangerous. Tomorrow held the promise of ultimate danger, eye-to-eye, hand-to-hand competition for life itself.

The men tried to rest and get some sleep on the bare ground under the open sky. The rains had stopped and the sky was clear and beautiful, but most of the troopers paid scant attention to the state of nature, even the bitter cold. Their thoughts were on tomorrow's action. They had waited along with Jack for this day, some with dread, some with eager anticipation, but all with the seriousness that only imminent death can bring.

As they sat among the heavy jungle brush and trees of their temporary home, Jack was anxious and contemplative. He had one hundred fifty men for whom he felt a tremendous weight of responsibility. He wanted to get all of them through this. Only thirteen of his men had been with Troop F from that day so long ago that the three young and eager Knight boys signed up for a year of adventure with the horse cavalry. Eight more were serving in other units of the 124th. Most of those remaining were now sergeants serving as platoon and squad leaders. Most of the original troopers had been re-assigned or discharged. Some had been killed in action in other war zones. Yet, he had

Curtis and Curtis' two brothers-in-law, Billy Ralph 'Scroney' Aaron and Grant 'Mickey' Crosland, both sergeants. Jack had formed a strong bond with all his troopers since his reassignment to the troop two months earlier.

"Curtis, I'm going to do my best to get us through this without getting hurt. You be careful and don't take any unnecessary chances." He didn't need to tell him why he was saying this. Curtis knew how he felt about his brothers and about Joyce and Glenda waiting for him to come home.

"OK. But Jack, I'm gonna do my part. I want to get my share of the Japs, too."

"Sure. We need to get some rest and sleep if we can. Tomorrow is going to be hell."

With the assignment of perimeter guards, the remaining troopers found the most comfortable spot to try to get a few hours sleep. Curtis couldn't get his mind off Joyce and Glenda. Curtis and Joyce had a love that was consuming and unconditional, a true romance. Glenda had been six months old when the 124th left Ft. Riley. She had her first birthday two days earlier. He fretted about their future and the void that he would feel if he lost one of them and transposed those feelings to them. He was deeply saddened and fearful that his young family could be shattered tomorrow, but his pledge to duty and devotion to his F Troop buddies reinforced his determination to help eliminate this nest of Japs and at the same time survive.

Jack was the last man in the troop to lie down. His thoughts, a jumble, raced on,

The stars are bright They look just like the ones back home That's silly why wouldn't they be This ground is hard I bet I don't sleep a wink all night That's O.K. I can kill Japs without sleep Hope Curtis gets some sleep so he can be on his toes tomorrow I'm glad he'll be with me He'll be right with me when we hit the Japs Have we planned enough Maybe there won't be any Japs on the hill They'll be there Maybe not too many Maybe the artillery will get most of them and they'll surrender before any of my guys are hurt Hope none of my guys get killed The Japs are crazy They'll come at us in a suicide charge before we get up the hill They'll kill half of us Is there a better way to do this Have squadron officers planned this attack right I wonder what Daddy would do I bet Mamma is praying right now I should of prayed more Lord, please help me get my men through this I will try to kill all of the Japs before they get a single F Trooper Do I have enough ammo I don't hear any noise from the troopers That's good or is it If they're sleeping I should hear some snoring Too quiet At least they're getting some rest I hope those artillery rounds are keeping the Japs awake Make them stay awake Equalizer What will happen to Mamma and Daddy if I don't make it What about the kids I haven't written in awhile Did I help any of them with my advice My butt and back aches I'll never get to sleep I hope to see the kids grow up We'll buy a good farm with a nice house All of us will go to college and get good jobs and save some money

and have kids I'll marry my little Mexican senorita I love her We'll have a heck of a homecoming when this is over Mamma won't know when to stop cooking and Daddy won't ever quit bragging on his kids Will he be proud of me June Bug will grow up to be beautiful and smart and I won't let her marry some sorry Joe-blow Bill will do O.K. He has the brains, but he's a little tender-hearted Gonna have to get tougher Don't have to worry about Nig being tough enough but he needs to get serious about school Wonder what's wrong with Smutter He'll probably go to the army before he graduates Maybe the army will knock some sense into him He's a good ball player Wonder if he'll be good enough for the pros I believe Herchel will make it through the war He has survived some of the worst part He should be safer on that battleship than the infantry I wonder where Loyd is Will he get a crack at the Japs We should have enough money saved between us to start that horse ranch We could make a go at that He's plenty smart and we know horses What will happen to Mamma and Daddy and the kids if I don't make it home Have I helped them enough I don't know if I could have done more Lord, forgive me if I didn't and help us all tomorrow I love these guys I hate those Japs Is that wrong If it is I can't help it Forgive me Lord, for I know what I will do This is war They jumped on us, the sneaky bastards Can't let 'em take over There's a time to kill That's what Mamma says the Bible says I know she's worried herself sick over the war She's tough She'll make it I wonder if Mrs. Broom is still making those dishpan banana puddings I wonder how many of the Garner boys will get through this war I bet Rowe is giving those Nazis hell They'll have them whipped before we do the Japs The German people have more sense than the Japs Not much more They let Hitler that crazy son-of-a-bitch lead them down a path of destruction We'll have to kill every damn Jap there is before they'll give up May take five more years We won't make it that long Might as well go for it tomorrow with everything I've got I won't get out of this thing alive anyway Can I get the job done Am I ready What are we gonna see on that hill I have good men Wish Curtis was still cooking No he's a good fighter and would volunteer to go with us anyway Just like Sergeant Doyle I've got good men They'll do o.k. Those Japs don't know what's coming at them tomorrow These guys have blood in their eyes Will all of them fight like men How do you know what you'll do in a battle if you've never been in combat I only know what I'm gonna do I'm gonna kill every damn Jap I see I need to lead by example Curtis will too Murphy can shoot out a gnat's eye He'll do O.K. and he's got the guts Pot-gutted Murphy What a name Good ball player Can move fast He's strong All of us are in shape We'll be o.k. This sky looks just like the one at home We used to sleep outside in the summer Cooler I'm a little cold I hear a few snores Not much movement Our guards haven't spotted any Japs They know better than to fart with F Troop They don't know better than to fart with F Troop They don't know about F Troop For all they know we could be B Company of some paratroop battalion Will the F Troopers make it through tomorrow They'll fight like hell Good troopers

Know how I'll lead They will follow Where He leads me I will follow Where He leads me I will follow Where He leads me I will follow, I'll go with Him with Him all the way.'

Sleep? A nod, no more.

Jack checked his watch and saw that it was nearing daylight and time to move. There had been little sleep. Jack & Curtis had had none. Thoughts of the day ahead had pushed aside any other thoughts and craving for sleep. Over and over in an endless string the troops contemplated what real combat would be like after practicing for so long.

The U.S.S. Maryland was still moored at Pearl Harbor getting repairs for the Kamikaze hit at Leyte Gulf. Loyd had sailed for Pelilieu to occupy the island and wait for further action.

Jack

CHAPTER 21

THE BATTLE OF KNIGHT'S HILL

JACK WAS FINALLY facing the enemy, men of a race he had been taught to hate. Yet his motivation superseded these feelings. He had been assigned the responsibility for the men of Troop F, for their safety, their success. It was his duty to fulfill that responsibility. His soul, mind and strength were drawn to a single thought. He would do his duty.

At first light on the morning of February 2, 1945 men began to stir quietly. Those who could do so, ate. They fitted their gear. They fastened their water canteens and squirreled away light rations and fastened hand grenades to their harness. They checked their weapons, making sure they were loaded and that they had a good supply of ammunition. Both Jack and Curtis were armed with a .30 caliber semi-automatic carbine and a shoulder holstered .45 caliber hand gun, with grenades hanging from their straps. By 0600 they were ready and on the line to attack. The three troops of Second Squadron had approximately 400 men ready for the attack. Unknown to them at the time, the Japanese had more men than they expected, about the same number as Second Squadron, in well fortified and concealed defensive positions.

Jack gathered his officers and non-coms and spoke quickly and quietly,

"O.K., F Troopers, we've been waiting a long time for this. You know what to do. Think about what we've been over and remember to do it. Let's do our best, but try to come back. If any of you guys don't go home, I won't either." Lt. Leo Tynan, forward observer from the 613th Artillery Battalion who had joined Merrill's Marauders in the battle for Myitiana, had asked to be assigned to Jack's troop, "Lt. Tynan, I want you to stay within three feet of me at all times. Let's go."

It was 0600. The troopers were in position. After twenty minutes of artillery fire on the suspected Jap positions, Second Squadron was ready to execute Major Jordan's battle plan. Troop F moved out at 0620. Jack and Curtis led the men at a quick pace down the hill toward the valley between their position and their objective. The troopers were in top physical condition. The men followed in a three hundred yard front, spaced five to ten yards apart. They were headed in a southerly direction to attack the northern segment of Loi Kang Ridge.

The area was covered with dense jungle and tall trees, which hampered movement. Widely spaced to minimize casualties from incoming Japanese mortars and artillery, the troops moved cautiously, looking for snipers in the trees. With these distractions and lack of a good field of vision, getting lost was a very real possibility. Curtis moved at a double time pace between leading elements of the platoons to keep them on the correct line of attack. His nighttime possum hunting experience in his youth helped prepare him for this. The platoon leaders couldn't see one another. As they pushed their way through the heavy growth, trying to stay abreast, they found at times small groups converged on narrow paths.

They had to cover 1500 yards as quickly as possible but all the obstacles slowed their progress. They reached the bottom of the hill with another 75 to 100 yards of slope to climb through thinning growth before reaching their objective. As they moved up the slope, two Japs on outpost duty, caught between the onslaught of this band of twentieth century knights and the steep slope behind, decided to report their intelligence to their commanders. It was too late. As they jumped to run, Jack spotted them and quickly took a knee and fired, killing both of them with several shots as quickly as he could pull the trigger of his carbine. The head of one exploded, the brains and blood gushing from the dismantled cranium as the troopers passed by.

The troopers by now were under heavier bombardment from Japanese artillery and mortar fire. Sergeant Wayne Doyle, the mess sergeant, had volunteered to fight to help build up the undermanned troop. He was one of the original F Troopers from Mineral Wells. He was serving as Jack's messenger and fellow warrior. He kept pace with his devoted leader. He would later pay the ultimate price. He was a true patriot.

As Jack and Curtis topped the hill, the northernmost part of the Loi Kang Ridge, Jack quickly surveyed his immediate surroundings of thick scrub brush and scattered trees, sparser than the first leg of their approach. It was 0710 hours. Drawing no fire and seeing no evidence of Japs he yelled to his men,

"There's nothing up here! Come on up!"

The men without hesitation topped the hill and breathed a short sigh of relief, but with some disappointment. Maybe the I & R unit had been mistaken and there were no Japs here. Jack ordered them to assume a defensive position and dig in. They had little time, even to locate positions for foxholes, for Jack and Curtis were on the move.

The troopers had been taught that the Jap's had a reputation of fighting to the death. The Mars Task Force G-2 had intercepted messages that indicated that the Japanese troops in this area had specific orders to do just that. With this in mind, the troopers dispersed to locate the ideal location for a defensive position to dig their foxholes. Jack and Curtis moved on. Their intuition, caution and training sent them to investigate the southwest slope of the hill. Surely there were more Japs on this hill. They were right. The Japanese had the advantage of cover and concealment. The Second Squadron had the advantage of cause and Jack and Curtis and Leo and Wayne and Willie and the Boys of Troop F. Curtis was about ten yards to Jack's left and a few yards ahead.

Curtis took cover behind a tree to search for Jap positions. He saw through a clearing a two-man Jap mortar crew running up the hill. He fired the first shots in the battle, hitting one of the Japs. The other one grabbed his buddy and dragged him behind some cover. Their objective would be delayed. A Jap bullet knocked bark from the tree a foot above Curtis' head. At that moment all hell broke loose. Jack had walked on down the hill and at the moment Curtis fired, he saw a Jap pillbox. Imitating his competitive ability to peg second base from his catcher's position and cut down the base thief, he rushed the pillbox and hurled a grenade into this pit of surprised sons of the sun. The blast eliminated another group of sons, brothers, and lovers. Would their families and friends ever know their fate? If they lived in Tokyo, Hiroshima or Nagasaki, probably never. These thoughts did not exist in Jack's mind. His men were beside and behind him, directing fire at every conceivable Jap position. He saw another pillbox and took it out. He was running as fast as he could among the Jap positions. The enemy in front, his men behind. Love and hate flowing from the same beating heart. Jack called out,

"Come on, there's a whole nest of 'em down here."

Lt. Tynan watched this furious assault and thought, ***What have I gotten myself into?***

Jack kept going and found himself caught in the crossfire of a horseshoe formation of pillboxes and foxholes. In a defensive position such as this, especially in the covered pillboxes, the Japs had two options today. They could hunker down and wait for death or they could come out and face a quick and certain death from the rapidly advancing force possessed by a desire to survive in the only way they knew; by a furious attack to get the Jap before the Jap got him. Both choices were made. Each choice produced the same result. It was face-to-face, hand-to-hand mortal combat with Japanese artillery and mortar fire and grenades exploding among them.

All of his cavalry training in a regiment whose motto was 'Golpeo Rapidamente'(We strike quickly.), all of his pent-up frustrations and his hatred of the Japs came into play. His love for his troops provided the extraordinary energy for the next few minutes of action. Self-preservation had flown from Jack at the first sight of the Jap, replaced by an overpowering burst of F Trooper preservation. En Masse, F Troopers had spilled over the crest of the hill to join Jack as he dashed from one pillbox to the next foxhole, firing and throwing grenades in a consuming effort to destroy the entire Jap contingent before his troopers came into harms way. As he ran for the next nest of Japs, Lt. Leo Tynan saw a Jap mortar man rise from a foxhole to shoot Jack in the back. Lt. Tynan, so close he only had to move his carbine a foot, at close range blew the side of the Jap's head away. Jack was hit by shrapnel from a Jap grenade. He yelled, "You little bastard!" He had sustained a non-vital flesh wound, a laceration on his forehead. Blood from the wound was running into his eyes. One eye was swollen shut. He ran out of grenades and ammo. He retreated a few steps to Lt. Leo Tynan for carbine ammunition and then to Sergeant Bill McQueary,

"Willie, give me some of your grenades."

Bill placed a few grenades in his helmet and he went back to the maelstrom. Love can't be qualified. He didn't think, 'Come on guys, if you want me to fight, you've got to fight.' The air was filled with bullets and, from both sides, grenade and mortar shrapnel and smoke. Jack, his vision acutely impaired, resumed his attack, exhorting his men when no exhortation was needed. As he resumed his attack, he ran past Lt. Tynan and uttered "I can't see!"

He tried to organize his men with arm motions, "Come on, we've got 'em now! Go on through! Go on through!"

Curtis had been searching the terrain for more of the enemy. The confusion and noise of battle makes it difficult to recall specific time spans, but only a short time later when Curtis heard Jack yell, "You little bastard!' he knew he was hit. He dropped to the prone position to reduce his silhouette and lessen the chances of getting hit and started toward Jack. He didn't know that a Jap was in a foxhole thirty feet in front and to his left. As he began to crawl toward Jack, the Jap stood up and fired into his chest. The steel-jacketed bullet from a .25 caliber rifle entered the left side of his rib cage a few inches below the armpit and, without exploding on impact, lodged just under the skin on the lower right side of his rib cage, piercing both lungs, but narrowly missing his heart.[He was later told that the heart had dropped slightly as he hit a horizontal position. The bullet had passed through an inch space between the heart and the spinal column, missing both.] He fell, paralyzed from the waist up.

Jack, now on the ground saw this. "Curtis, are you hurt?"

"Yes!" someone yelled.

"Go on back. Somebody get Curtis back," Jack yelled.

Curtis could yell. He was giving commands to the men to move forward and to help Jack. Sergeant 'Red' Wilson and others dragged Curtis to cover. Sergeant Wilson gave Curtis a canteen of water and a pill to slow the bleeding. He prayed with Curtis,

"Lord, please help Curtis. Put your hand on him and slow the bleeding. And Lord, forgive the Japs, for they know not what they do."

[Curtis first told me about this prayer fifty-seven years after the fact.]

Sergeant Kim Hill jumped in a foxhole that he had just grenaded. To his surprise, he landed next to a very live Jap who immediately tried to stab him with a sword. He grabbed the Jap's sword with one hand and stabbed him with his other hand. Clyde Stockton came face to face with one of the Japs who chose to leave his nest. He fought him hand to hand until he prevailed.

The blood had flooded Curtis' rib cage and the pain became unbearable. Five minutes after he was hit, he passed out from the pain and was evacuated. Jack propped himself on one elbow, blood flowing over his face and into his eyes. Thoughts of Joyce and Glenda without Curtis flashed through his mind, one of his worst fears. He urged the men forward, his functional arm directing the action. With the last surge of strength in his body, borne of fury, he got to his knees and tried to reach one more pillbox. Without quickness and mobility, he became an easy target. He was hit in the chest by a burst from a Jap machine gun. He lurched forward, his last spark of strength devoted to duty.

'What is this? This stuff tastes like blood and dirt. Why am I eating blood and dirt? Someone is standing on my chest. Men are running and guns are shooting... gunshots. I need to get to my men. Where are my legs? I can't find my legs. Where am I? Willie, Curtis...Daddy......Mamma.........When the trumpet of the Lord shall sound, and time shall be no more, And the morning breaks, eternal, bright and fair; When the saved on earth shall gather over on the other shore, and the roll is called up yonder, I'll be there.

Oh, God!
Our Jack is dead
My Jack is killed
The god of my childness is gone
He fought, was just
And he died
He vanquished the foe
But he is dead
He had divine immunity
Yet he is slain
Oh, speak wisdom to me Solomon
'Set me as a seal upon thine arm:
for love is strong as death...
Many waters cannot quench love,
Neither can the floods drown it'
And Jesus said,
'Greater love hath no man than this,
that a man lay down his life for his friends.'
Why!
Why did he die?
Why did he sacrifice,
Why did Roy sacrifice,
Why do they sacrifice still.
We know.
It is love.
Amen.

Kachin Rangers

Kachin Scouts

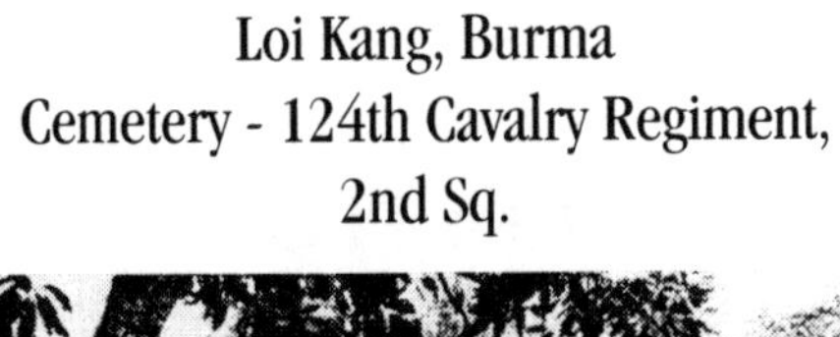

Loi Kang, Burma
Cemetery - 124th Cavalry Regiment, 2nd Sq.

Kenneth Payton

His task was to place dog tags between the teeth of F Troopers killed on Knight's Hill

CHAPTER 22

BEYOND THE END

PVT. ANTHONY WHITAKER, seeing Jack lunge to the ground, rushed forward. He was armed with a bazooka and fired into the pillbox that Jack was attacking. The round failed to explode. He rushed the pillbox, threw a grenade into the hole, killing the men inside. Tragically, he was killed in the attack.

The battle was over for the Knight brothers in no more than twenty minutes. Yet, the tone of the struggle had been set. In S-2 reports of the 5332d Brigade the action was described as savage. As Troops G and then E joined in the struggle, the fighting raged into the afternoon. Troop F had only two other officers and both of them were wounded early. Lt. Gotwald, Third Platoon leader was shot in the shoulder. Lt. Bayless, Second Platoon leader had his mandible virtually destroyed. Lt. Leo Tynan, the artillery observer, took command of Troop F and formed a defensive position and led the troop in the final assault on the Jap emplacements. Lt. Tynan had already earned the coveted Combat Infantry Badge for his experience with the Merrill's Marauders. What he did that day supports my belief that much of our success in World War II was the result of almost flawless transition in command on the battle field. Trained as an artillery officer, Lt. Tynan took over a combat infantry operation and distinguished himself as a leader of men. He was awarded the Silver Star for his heroic and sacrificial performance in the Battle of Knight's Hill. On March 13, 1945, Major General Vernon Evans by command of Lieutenant General Sultan, wrote the following citation in awarding the Silver Star.

"SECOND LIEUTENANT LEO C. TYNAN, 0530147, field artillery, United States Army. For gallantry in action near Hpa-Ping, Burma on 2 February 1945. While serving as forward artillery observer, Lieutenant TYNAN noticed that all the officers of one unit had been wounded or killed as a result of heavy enemy fire, and that as a result, the unit was without control. He assumed command of the unit, reorganized it, and continued the attack, functioning at the same time as artillery observer and unit commander. The initiative and devotion above and beyond the call of duty displayed by Lieutenant TYNAN reflect great credit upon himself and exemplify the highest traditions of the Armed Forces of the United States."

Excerpts from John Randolph's 'Marsmen in Burma' will describe the remaining action and thoughts for the day.

"When the tumult of the battle was over, and the excitement of making evacuations, digging-in, avoiding sniper fire, and exchanging stories died down, the most oft-repeated words were: "Jack Knight is dead!" "Those bastards got Jack Knight!'

"He was a trooper's trooper, loved and admired by all. They knew he would do all that he did. They were sorry that he tried to do so much alone."

..."Part of the artillery support given G Troop consisted of phosphorus shells. The breeze was blowing toward E's and F's positions on G's right. Japs, trying to get away from the smoke and fire, were pushing into E's and F's newly won and barely held perimeter just as both of them were taking stock of their positions and were realizing their inability to hold them against any forceful counterattack because of manpower shortage. But as the Nips came in, they were mowed down by our men. Fortunately, it was not a counterattack, just a few escaping Nips who did not know our attack had been extended on such a long line.'

"Noon came, and with it, word that L Troop was jumping off in an attack. As if there wasn't enough hell going on!'

"Evacuations could not be made fast enough. F Troop was calling over and over for ammunition, especially grenades. Sergeant Whatley hurried over to the aid station near the squadron CP and collected arms, ammunition, and grenades to send up by returning evacuation teams.'

"Deceptive terrain again brought its hardships, both in the attack and in the establishment of supply lines. It was a long way through thick growth by indistinct trails from supply dumps to the line. Mules carried ammunition to the base of the hill, but the job of making contact with the troops and hauling it up the steep side was another problem. It finally got there, but not until after at least one troop had threatened to withdraw if no ammo or support could be sent.'

"At the CP they were trying frantically to collect men to fill in the gaps. Capt. Andrew J. Kaelin, CO of Regimental Headquarters Troop, was called on to find the Chinese company which was somewhere in reserve and take it up to fill the gaps. He was having trouble, first in finding it, and then in getting the results he wanted through interpreters.'

"Major Jordan, from the OP, contacted all three troops on the radio and ordered them to hold on. He told them that Colonel Osborne was confident that, if they could hold through the night, the Chinese would come in to cut the road farther south in the morning. The troops were not thinking of tonight, much less tomorrow. They were damned apprehensive about the immediate present. E and F lost contact and for a few tense moments each thought that the other had been forced to fall back. E had advanced its line and needed connecting support.'

"It was 1310 before all troops agreed that the situation was well in hand. At 1342, Lt. Ben Fuelberg, CO of E Troop, called to the rear for his first sergeant, Arnold Winkleman, to bring up food and water, a good sign that things were settling down.'

"Calls for medicos and litter bearers continued until late in the afternoon. At 1350 it was reported that 53 men had been evacuated to the portable hospital. Many of these had already been flown out. The aid station had some 15 wounded under treatment or awaiting evacuation or a return to the front, with more steadily flowing in.'

"Evacuation was exceedingly difficult. It took a long haul through sniper-infested woods to get men as far as the aid station. I saw S/Sgt. Harold S. Odde from Thermopolis, Wyo., who had come down as a pinch-hitting litter bearer from F Troop, head back to the front. In a few minutes I saw him at the aid station on a stretcher with a hole in his leg. He was one of many to meet the same fate, or worse.'

"Capt. James M. Huey, Squadron Surgeon, of Louisville, Ky., was doing a Herculean job with his assistants at the aid station. This was one place where there was virtually no confusion, despite the pressure of activity."

..."The 49th Portable Hospital became the scene of well organized yet feverish activity which did not let up for 72 hours. As many as four operating tables, on which were performed miracles of field surgery, were in use at the same time. Doctors and technicians were sent away from their work individually, as they could be spared, to catch a well-earned short nap, and then they were back on the job again.'

"During our days of combat, this one small unit performed 27 major and 116 minor operations with a mortality rate of 1.7%.'

"At 1415, Lt. Julian H. Ward was finally able to round up 24 men and take them to the front to fill gaps. It was after 1500 hours when Captain Kaelin was finally able to get his Chinese where they were needed.'

"The battle was won.'

"Pfc. Ernest H. Barkley, a veteran from Merrill's Marauders, said, 'I went through five attacks in the battle for Myitkyina. This was my sixth attack, and the biggest and toughest I've ever been in. I don't want another one like it.'

"Colonel Osborne said, 'In over four years of combat I have seen many officers fight and die for their country, but the actions of Lieutenant Knight in leading his troop against a strong enemy will always remain as the finest example of American courage, valor, and leadership of any officer I have had under my command. It is officers of Lieutenant Knight's caliber, and the troops that follow that kind of leadership, who are winning the war—not colonels and generals.' This is no idle tribute. Colonel Osborne had been through the siege of Bataan as well as the Merrill's Marauders' battles."

"Lieutenant Cornwall, who later became CO of F Troop, was with G Troop in the attack. His reaction, as recorded in some notes jotted down for his personal use, was as follows: "The only positive thing I will remember about the attack was its boldness...They shot it out man to man and the Americans won... seldom have braver men died than the ones I saw on the slopes of this vicious hill. We lost terribly, but we gained the objective, and we killed Japs.'"

"Yes, we killed Japs. Over 200 were buried on the hill."

"In a separate cemetery for 2nd Squadron, the bodies of 22 brave troopers were buried. In a bamboo grove with a bamboo fence around it, 22 white crosses mark their graves. Each body was wrapped in a white parachute and covered with an army blanket."

..."The hill was officially named 'Knight's Hill,' the cemetery 'Knight's Cemetery.' The Battle of Knight's Hill was over, but would never be forgotten."

The naming of Knight's Hill was decreed by Lord Louis Mountbatten, Supreme Commander of the China-Burma-India Theater. This would not be the only thing he did to honor Jack.

To complete John Randolph's chronicle of the Mars Task Force a final entry will be made,

"The 124th buried 516 Japs and the 475th buried 157 for a total of 673 known dead. No official estimate of those who were killed but whose bodies we did not bury was recorded, but there is ample reason to believe this total was high, too. Four prisoners were taken: two by each of the regiments. We had destroyed or otherwise prevented the removal of large quantities of supplies stored in the vicinity of Namhpakka."

..."The price paid in American casualties was evidence of a score very much in our favor, but it was high. Marsmen numbering 122 were killed in action or died of wounds. Sixty-four were cavalrymen. Five were men of the 613th Field Artillery. Fifty-three were from the 475th Combat Team. One man was missing in action."

Interviews with F Troopers indicated that nine F Troopers were killed. While interviewing Kenneth Payton and his son-in-law, Dr. Robert Orr, in April 2004, he said that his awful task after the battle was to place the dog tags of those killed between their teeth.

Curtis was carried by some of the F Troopers to the squadron medical aid station. A guard was posted near his cot to see that no one mentioned Jack's death. In a conversation with Dr. J.M. Hughey at the reunion of F Troopers in April, 2000, I discovered for the first time about his stay at the aid station. Dr. Hughey, examined Curtis as he was brought in. He told me that Curtis had no measurable blood pressure and hardly any blood in his system. He gave Curtis large doses of antibiotics and two quarts of plasma.

Curtis told us soon after arriving home in June, 1945 that he was semi-conscious and late that day he heard the doctor tell a medic that he wouldn't make it through the night. He said his reaction was thinking, 'I'll show you.'

He was moved to a field hospital and later to India to recover sufficiently to be sent home four months later. In September he was informed that he had been awarded a Silver Star Medal for valor, along with his Purple Heart, for his heroism on Knight's Hill that day.

Jack was laid to rest along with sixty-three other warriors of the 124th Cavalry Regiment on a small open spot in a valley near the battle site.

Mamma continued to write Jack and Curtis through February. I know that she wrote more, but I have four letters that have survived. I will quote them to show the mood at home.

Sunday P.M., Feb. 4, 45:

"Dearest Jack,

"How are you? We are well & sure hope you & Curtis well & safe.'

"R.C. came home again today. Will have to go back this eve. There were church & we went & just got home. So we will take R.C. back about 3:30 so he can get there in time to milk. He doesn't know yet if he will be reclassified or not, but thinks he will but I kinda doubt it. Sure hope not.'

"Well I'm still holding my job. Dad & I will draw over $200 our next payday & its good pass time but when Joyce comes back I guess I'll quit & we sure would like for them to come back.'

"We've been hearing news from Burma, but none the last 2 or 3 days. Sure hope you are all getting a rest.'

"Jack when you write say something about Red Wilson so I can tell his mother & if he can write tell him to write her for she wants to hear from him.'

"We had 2 letters from Herchel this last week. He was feeling fine. Didn't hear from Loyd this week. We got your & Curtis letters of the 13 of Jan. Was so good to hear from both of you & will be glad to hear from you again.'

"We just heard one of the prisoners talk that was released by the rangers on Luzon Island. I guess you heard about it. There were a bunch of rangers went in & captured & released 510 of our prisoners. I know that was a happy time for them.'

"I hope you boys get plenty to eat. I want to send you all something as often as we can.'

"Did I write you about getting some of your bonds. We got Sept & Oct. bonds $400 dollars worth. And for fear you didn't get my letters I wrote about the money we got $270 & $40.'

"Kenneth Knight is home on his last furlough before going across or he thinks so. Dad & kids went to grannies this AM. & saw him.'

"Connie Marie said she heard from Rowe last wk. He was in Holland. She is still working at the Camp. I'm learning lots about soldiers cloths, how to classify. We fix the best ones to send over seas & the next for the soldiers here & the

X's for the prisoners. I can size & mark now it is easy. When I'm there a month or when I get a full pay I get a 3c raise, then I don't know, but I guess I'm as well off or better than I'd be here.'

"Well Dad & the kids has gone to take R.C. back. June & I decided to stay here & write.'

"I'd better close & write the other kids.'

"I hear the news just now, said the Russians are 40 mi. from Berline. I hope it ends in Berline, but afraid not. Seems like we'll soon have the Philippines back. I'll be glad too when they get into China & finish that part up.'

"Well Aunt Matties folks came this eve. & stayed awhile. They are about as usual. They havn't heard from Johnnie since he went across & the last they got from Norris was written the 1st of Jan. They hear from Fred. He's in France I believe.'

"Well be real careful & write when you can. We love you all.

Mother & all."

Thurs., Feb. 8, 45

"Dearest Jack,

"I'm at work. I can write at rest periods. We have one at 10:15 AM & one at 3:30 PM & 1 hr. at noon. So I have more time here to write than at home. But we are doing OK with everything I guess.'

"Well this is noon & I've fiddled around trying to remember how to tat. We crochete, tat & write and a little of everthing. You may not remember what tat is. It is made with thread. Can make lace & things like crochete.

"There was a picture in the M. Wells index from Burma, some boys loading a truck, but they didn't any of them look like any of you, except one Dad said looked like Red Wilson. His mother thought it was him too.'

"Miss Lou Sullivan & Mrs. Bolton works across the St. from us with a whole bunch of other women. The weeks pass off quicker since I started working, but I think when Joyce come back I will quit & make a garden & raise some chickens.'

"We had a notice from the gov. written the 19th of Jan stating you were in good health. Got Curtisses about 2 days ago.'

"We had a fire across the hiway Tue AM. The little shack burned and most of the things. The woman got out a few things. It's a soldiers wife. So she & her brother in law is staying with us till she gets a place to stay. She was taking my clothes to town today to wash for me.'

"Well this is night and we just got the dishes washed. Mr & Mrs. West & his brother are here, but he will have to go back to camp tonight.'

"We had an announcment today about H.D. & Doras new baby girl, name Esta Ruth.'

"We don't have much news to write, tho I guess anything would be news to you boys.

"Seem like the war news all sound favorable now. We hear news from Burma & it sounds good, tho we hate to know you boys are in it.'

"Well, I'll have to close & go to bed.'

"Write soon as possible. We love you bushels.'

Mother & all"

Unknown to us, Jack and Curtis had had their war. At about this time we had our little part of the war literally in our own back yard. Across US Highway 180 to the south lived our old neighbors from our last farm, the O.T. Simms family. They had several daughters but one son my age, O.T., Jr. While Mamma was working we spent a lot of time at the Simms' waiting for our parents to get home. We roamed the gullies and woods behind their house or played ball in their front yard.

During one of these visits to the Simms' I was in their front yard with some of each brood playing catch. Our house across the fallow field was in full view. We heard our dog, Bruno raising an awful fuss. We had lost Ted to poison bait put out by a goat rancher to eliminate the canine invaders that were killing his stock. As we looked, we saw nothing but Bruno barking fiercely at the front door. This was puzzling, but his actions were explained a moment later when a man in coveralls ran out the front door carrying some clothes. Bruno stayed on his heels as he disappeared behind the house, running into the woods. It was late afternoon and Mamma and Daddy were soon to be home from work. This was during the few weeks that Mamma worked for wages; the only time in her life she did this. We ran into the house and, based on our knowledge that war prisoners were housed at Camp Wolters, yelled,

"Mrs. Simms, a German prisoner broke in our house and Bruno ran him off!"

"Did you see him?"

"Yeah! We saw him and he was running out of the house and Bruno was biting him."

Neither home had telephones, but the gate to Camp Wolters was only five miles to the west. Mrs. Simms told us to stop a car on the highway and ask them to report our discovery to the M.P.'s at the gate. We were quickly successful, but Mrs. Simms wouldn't let us go home until the M.P.'s arrived. That didn't take long. In less than thirty minutes we had several army vehicles and M.P.'s all around the house. They organized and began the search.

Mamma and Daddy arrived home shortly after with great consternation. In minutes, Daddy and my older brothers who were still at home, R.C. and Roy, were organized into a small, well-armed band making a formidable search party. Daddy assigned weapons.

"R.C., you get the double barrel twelve gauge. Junior, take the twenty-two and those hollow points. R.C. get all the buck shot you can find. I'll take the pump."

All of this frantic action made Mamma nervous.

"Roy, you better let the M.P.'s do this. You or the boys might get hurt."

Daddy was not going to be denied this role. He had missed The Great War because of his foot injury and the birth of Jack. He would never have another chance to face the enemy.

"It'll be O.K. We know these woods better than he does and he probly don't have a gun anyway."

Mamma knew to quit arguing. She would really have something to worry about with a German in the woods. Daddy quickly assembled the troops and commanded,

"O.K. boys, let's go. He has a head start, but we might find him hiding from the M.P.s in the brush."

I had no weapon nor assignment, but I wanted to go. Quietly, I walked around the house out of sight of Mamma and as the warriors headed out I surreptitiously followed. I was following the concept that I would learn as an adult to be the 'It's better to apologize later than to ask permission' strategy. When I was discovered, we were well into the woods and Daddy reluctantly allowed me to go along. He had eagle's eyes and knew that I had inherited at least some of his keen eyesight. He could use an unarmed scout.

"O.K. Bill, you stay right by me and help me look."

We crossed a creek along which grew post oaks, live oaks and thick brush and headed to the north and east. We fanned out about forty to fifty yards apart and scouted the woods in a half-mile radius of the house until the sun was brushing the horizon. Reluctantly, Daddy called off the search.

"Well boys, we've done all we can. He's probly half way to Weatherford by now. We can't see good enough in this brush now anyway. Let's go to the house and see if we can find out what he took."

On the way back we passed the soft stone formations along the eroded banks of the creek where Junior had sculpted several generic faces with his improvised tools of spoons and screwdriver. We were amazed at his talent. He had a great ability to work with his hands. He had shown this before by carving miniature farm equipment and airplanes with a pocketknife. I had tried to emulate his work, but my efforts were always askew and out of proportion.

As it turned out, we weren't looking for a German at all. One of the Camp Wolters trainees had gotten into trouble and was in the guardhouse. He had slipped away from a work detail and was on his way home to East Texas. He needed civilian clothes. After an inventory, Mamma and Daddy determined that he had taken only a change of clothes and Daddy's felt fedora. He might have taken some food, but it was hard to tell. The Knight Stalkers gave way to the M.P.s and resumed normal activities.

The next day the M.P.s were searching an area east of our house. They were on a ranch that had rolling hills and open prairies dotted frequently by groves of live oak. These groves usually had thick brush undergrowth. As a contingent of M.P.s passed one of these groves, the hapless escapee jumped up and began to run.

He immediately stopped short as a simple command to halt rang out. During the interrogation, it was learned that he jumped, not to escape the M.P.s, but from the intrusion of a harmless grass snake that got just a bit too close. We were safe again. That boy must have lived in one of those East Texas towns with no country connections if he couldn't identify the snake as non-poisonous.

Weeks later Daddy was called to testify that the clothes were his. The poor misfit received a sentence of fifteen years in federal prison for escape and burglary of a habitation, thus ending the homefolks only armed participation in the war.

Friday, 2-16-45

"Dearest Jack,

"How are you & Curtis by now. Hope you are both well & healthy & having plenty to eat. Looks like I wont ever get you another box mailed. Wish I could just drop you a box of stuff ever day. Just wait when you boys get home. I'll cook all the time or do all I can to—well I'd better hush.'

"We haven't had any news from the Mars Task Force in the last few days. Hope everthing is quiet over there and you are all getting to rest.'

"Joyce said she got a letter from a boy in Troop F telling her Curtis was OK, but couldn't write but she didn't say when it was written. So I suppose by now you all have heard or had a letter about Joyce going back to Poteet. We are ready for her to come back tho I don't know just when she will be back. We're going to keep them when they do come back.'

"Did I tell you R.C. is in 1A now. He is home. He said he might work in the camp till he has to go. He hasn't been examined yet.'

"I'm still on my job, been working a month yesterday, but don't know how soon I might quit, tho I still don't work hard.'

"We had a stork shower today for Sallie Howard. She is quiting tomorrow. You know Pat is 10 years old. That seems to be contagious around here. I heard Stokes & Nannie is in the same path, but it might be a mistake. We tease Mrs. Hobson & told her she would be next. She is 52 years old. She said she wouldn't care but she's pretty windy. She & Bernice Cowley keeps the static going over there all the time.'

"June is sitting here wanting me to go to bed with her, so I'll close for now. Be real careful. Give Curtis my love. Tell him I write him next & soon.'

"All our love & prayers.

Mother & all"

Tuesday, 2-20-45

"Dearest Jack,

"How are you? We are well & hope you & Curtis OK and doing fine.'

"I'm at the camp & its noon. I just went over to see Mrs. Wilson. She said she heard from J.B. yesterday, written the 6th. So mabe I will have a letter from you & Curtis when I get home. Sure hope so.'

"We had a letter from Loyd yesterday & Sat. He didn't tell where he was, but said there wasn't anything to worry about. We had a letter from Herchel Sat. He was OK & still at Pearl H. I think. Hope he get to stay there for the duration.'

"The news sounds good ever where. Mabe it won't last too much longer, tho one day is too long.'

"R. C. is home now. He could stay out I think but he doesn't want to seem like.'

"We are putting out lots of cloths now, tho it isn't hard. They arrange for the work to be light on the women. I fold, count, and keep books. We put out from 5 to 8 hundred per day, but sometime part of that is canteen covers & first aid pouches & little knit wool caps.'

"I showed a captin here in this building your picture & he said he's nice looking. It is real C.S. stuff. He ment you & what you had on. C.S. is cloths we send overseas.'

"Well this is later, at home. We didn't hear from you & Curtis today, but expect to soon.'

"I don't have any news to write.'

"We still have rainey weather. Can't find a day to wash. I'll be glad when spring comes.'

"Do you all get any news now. It all sound like it in our favor.'

"We haven't had any news from the Mars Task Force in several days. So we are hoping you all are getting a rest.'

"I had a card from Elizabeth Crothwate today. Said Bill Tucker is in France. Wanted Curtis address.'

"Paul [Watkins] said they had heard from all their boys the last few days. They are all in Europe.'

"Well I've got to write Herchel, so will close and hope to do better next time.'

"Be real careful. We'll be thinking of you all. All our love and best wishes.'

"May God bless & take care of you all'

Love,

Mother & all"

June added a note.

"Dear Jack,

"What are you doing. I am fine and hope you are too.'

"I went to school today. We are having a rythim band. I play the drum. I missed two words in spelling today. She gave out watch and I thought she said wash and she gave out bake and I forgot to put the e on."

Love June Knight"

The anxiety around our house continued to build until a few days later when we got the bad news. Several weeks had passed since any of the parents and wives had received a letter from Burma. In mid-February some began to receive mail. None was forthcoming from Curtis or Jack. Unknown to Mamma at this time, all of her letters from December 1, 1944 had not reached Jack. They were later returned, all stamped 'deceased.' Mamma and Daddy had been worried all along, but knowing Jack's unwavering habit of writing often, they were now wrought with anxiety. Our fears were justified when, on February 24th, the taxi drove up the lane to our house to deliver the telegram about Jack. Soon after, we learned from Joyce and then by telegram of Curtis' wound.

The taxi driver stopped short of the front yard. It was hard to tell where the driveway ended and the yard began. All of us had migrated to the front yard. We knew. He gave Daddy the telegram, stood for a brief moment, got back in his vehicle and drove away. This was probably not his only mission of despair.

Daddy read the telegram silently. He placed his arm around Mamma and laid his head on her head while she read. He didn't need to read it to the rest of us. "It's about Jack." And Mamma's silent tears finalized the message.

All the years of hope, prayers and childish dreams of invincibility crashed. Jack was dead; killed in Burma by a Japanese soldier who was trying to survive, but of course, he didn't. Few of the Japanese in those pillboxes in sight of F Troopers did. Jack's men saw to that.

We entered a period of grief with a focus on Curtis' condition and the welfare of Loyd and Herchel. Continual distress was the emotional mode of the time. We began to hear, through the relatives of the other F Troopers, about the events of February 2, but the most definitive description came through the newspapers with an Associated Press release on February 25. We knew only then what had happened.

The U.S.S. Maryland would soon be seaworthy and on its way back to the South Pacific to take part in the invasion of Okinawa. The Maryland participated in the bombardment of the island for several days. On April 17 at 1849 hours a suicide plane avoided the AA fire and crashed into the ship on top of Turret Three. Herchel told me fifty-five years later that his experience at Leyte Gulf and the first few days at Okinawa taught him to listen carefully to gunfire. He was in the compartment next to his own. He and two other sailors were waiting for their shift on deck as signalmen. Herchel heard the gunfire changing from high caliber to lower caliber AA, meaning only one thing. A Jap airplane was coming at their ship. He told his companions to dive into a depression on the deck. They made it an instant before the plane hit his compartment. None were injured. The Maryland had to return to Pearl Harbor for repairs. For the USS Maryland the war was over. After repairs were completed in late August, she began shuttling troops from Pearl Harbor to San Diego, San Francisco and Seattle for the next few months. Herchel was discharged in the spring of 1946. (USS Maryland deck log)

The Secretary of War, Henry L. Stinson, on May 3, 1945 signed the proclamation posthumously awarding Jack the Medal of Honor for his intrepidity in the battle. The news was released and on May 10, reporters and a photographer from The Fort Worth Star-Telegram came to Millsap for the home story. They interviewed and photographed Mamma and Daddy along with Joyce and Glenda with a photograph of Curtis.

They then showed up, unbeknownst to us, at the school. When the superintendent called over the intercom (We didn't have those at Garner.) for us to come to the office, I responded with great trepidation. I hadn't to my knowledge violated any rules. I was a good kid or at least tried hard to cultivate that image. Roy, June and I sidled into the office, wide-eyed and unsure.

After a while with all the commotion and picture taking, I first realized that Jack had done something extraordinary. I was proud of him, but deeply saddened. I had already made a vow to myself that I would think about him at least once a day for the rest of my life.

Infantry Day was observed on June 15, 1945 at Camp Wolters. On that day, Maj. Gen. Bruce Magruder, commanding officer of Camp Wolters Infantry Replacement Training Center presented Jack's Medal of Honor to Daddy.

To gain more insight of the time and the effects of war, I asked my baby sister, June to summarize some of her memories of World War II. She was five years old when the war started. Her memories about Jack and the war:

"Bill, some of my memories may not be entirely accurate. I think sometimes our memories of what we witnessed and what we've heard about kind of merge. Anyhow, I do believe most of these are pretty accurate. If you recognize something that doesn't seem right, just forget it.'

"My first recollection of the war was the day the Japanese bombed Pearl Harbor. I remember that it was Sunday and Daddy was sitting at the kitchen table listening to the news. (Did he hear it at the Garner store or did he hear it on the radio – I don't remember.), But I do remember that to me it seemed a very serious, grave situation. I know Mama and Daddy were both upset and very worried.'

"I remember going with Mama to see the boys at Fort Ringold. I think I was six. We rode with Aubrey and Mary Dack [Wallace]. I got car sick on the way, and I thought we would never get there, it was so far, and I had never been farther away than Weatherford. I remember very well Jack paddling me because I was being a brat, I'm sure, and me going out of Curtis and Joyce's house, and laying my head on a rock fence and crying. Later, they made me pose for a picture, acting like I was crying.'

"I remember having to move from our house on the lake to the tile house because of Wolters and the Government taking up all of that land around Wolters for training area.'

"I remember Daddy going to work at Wolters, and Mama worked for awhile, I believe in the laundry. I know we had more money to buy things than normal, but I remember Mama saying she would rather have things the way they were than having jobs because of the war. In fact, there was a woman, I believe, who worked where she worked that was saying she was glad we were in a war, because it meant more money, and she didn't care if it was ever over. If I remember right I believe she and maybe her husband were fired because of that statement.'

"Also, I remember the rationing, and the books of coupons. I thought they were the same as money, until it was explained to me what they were.'

"I remember the last time Jack was home on leave before he went overseas. It was evening, and I believe we took him to Mineral Wells to catch a bus, or maybe the train. I had gone to sleep, and they were waking me up to tell Jack goodbye, and I couldn't get awake enough to tell him bye. I thought about that a lot over the years.'

"I remember Mama praying a lot for the boys- and their safety, and when Jack was killed, how she didn't think she had enough faith, or had not prayed enough for them.'

"I will always remember the Sunday that someone came and told Daddy that he had a telephone call at the store and he went and when he came, he said it was the newspaper (Fort Worth Star-Telegram?), saying that Jack had been in a battle and was a hero and would be awarded the Congressional Medal of Honor, but the taxi driver brought the telegram and we learned about Jack and about Curtis being wounded. I ran to the back bedroom by myself and jumped in the middle of the bed and was crying, but there were no tears. It seemed like the end of the world for us, and I wondered how Mama would survive it. Then all the people started coming in, friends and relatives from Young County, and up that way. We missed a week of school, and I will never forget going back to school, and the teacher (but I don't remember which teacher) asking me where I had been, and I said my brother was killed. She said your brother was here? I repeated it, and she acted like it was nothing. I thought she must have been crazy, because I just thought everyone must know about it, because it was such a tragedy for us.'

"I also remember when the prisoner broke into our house and it shook everyone up. I thought it was a German prisoner, because they kept some at Wolters, then we found out it was one of our men, being held for something he had done.'

"I remember the day they presented Daddy with the Medal of Honor, and how they picked us up in military vehicles, and that Curtis had come home that day. One thing that embarrassed me terribly was that it was hot, the windows were down, and I had my hand sticking out of the window and lost my bracelet Curtis had brought me. I think it was a CBI bracelet. The driver had to turn around and go back and get the bracelet. (This was when we were leaving Wolters to go home.) The bracelet had been run over, but I was just as proud of it and wore it for a long time.'

"I remember going to the movies once in awhile and seeing the newsreels of all the fighting and being scared. It seemed to me then, that there was a kind of dark cloud over us as long as the war was going on. I can remember riding the bus to town sometimes. I guess it was because of gasoline rationing.'

"I remember when we bought the place in Millsap with Jack's money. It was hard for Mama to enjoy it because of Jack's not being there, and it was money he had saved. It was just very bittersweet.'

"I remember the day the war was over. I think it was August. We had gone to Fort Worth with Joyce and Curtis and went to a movie, downtown Fort Worth. When we came out of the theater, the place was covered with people and confetti was coming down on us everywhere. We found out the war was over, and there was dancing and hollering in the streets. Everyone was happy. I don't remember, Bill if you were with us or not, but I know Roy was because he had pinched me for some reason and I was crying, quietly, and standing on my knees looking out the back window. All of a sudden I saw sparks flying behind the car, and we had lost a wheel. We had to stop, but I really don't remember much of the rest.'

"This is a little more than about the war, but those are things that stand out in my memory, and I'm sure there are other memories, but they're hiding there somewhere. Oh yes, one thing, I don't know if I ever mentioned this to anyone or not, but for years, (many years), I would be thinking about Jack, and wish it was all a dream and I could wake up and it would be before he was killed and things would not have happened as they did. That happened many times."

My family has had its share of war. We hope that it will end, but it won't.

WAR

War is hell and
It is a ravenous beast that kills and
Absurd
Must we give our sons, our brave
To quell the tyrannical ravens of
Hate, greed, megalomania,
Religious fanaticism and intolerance
Is a blood sacrifice
The only way
To cleanse the world of
Evil
History, the great teacher tells us
Appeasement is the lifeblood of evil
The vision is clear
Stand up, go forth with courage
Take up the shield
Wield the sword,
Save the children and
Shed a tear

The End

Daddy receiving Jack's Medal of Honor from Maj. Gen. Bruce Magruder assisted by Captain Courtland Kram

Photos this page: Courtesy Fort Worth Star Telegram Photograph Collection, The University of Texas at Arlington Libraries, Arlington, Texas

Back: Daddy with Medal, Mamma, Joyce, Glenda, Curtis
Front: Bill, June, Roy, Jr.

EPILOGUE

Troop F

Before the battle of Knight's Hill was completely over, orders were being cut for the next movement of the Mars Task Force. The following orders were given at noon on February 2.

"Mission: To drive the enemy south of LASHIO; thereby securing the land and air routes to CHINA. New First Army, less one Bn 75mm Howitzers attached to 50th Div: will advance Southward along general axis of the BURMA ROAD on LASHIO.

5332d Brigade (Prov), less 1st Chinese Regt, Sep, upon being passed through by New First Army, will concentrate in the NAMHPAKKA-HOSI area in NCAC General Reserve, prepared for prompt advance.'

"CT 475: Upon being passed through by New First Army, will concentrate in present area prepared to move in any direction ordered by CG, 5332d Brigade (Prov), to strike and destroy the enemy in its zone of action."

Late on February 2, an operational summary mentioned that A and B Troops of Combat Team 124 repulsed enemy attacks during the night and that 2d Squadron moved to Mong Noi. Regimental CP changed location.

On February 3 Lt. Col. G. T. Laughlin, Regimental S-3 wrote the following for February 2.

"2d Sqdrn. Supported by 613 FA. Attacked high ground at...at 0620 hrs. 2d Feb. F Troop succeeded in getting on high ground by 0710 hrs. Bitter fighting followed and E Troop (in reserve) was committed and reached the high ground by 0800 hrs. Mop up action and close in fighting followed until G Troup reached high ground at 1100 hrs. Ground was then consolidated with assistance from one company of 89 Regt. Approx 200 Japs killed in this action. Our casualties were approx 90 wounded, 30 killed. 3d Sqdrn attacked high ground at...1230 hrs. 2 Feb. L Troop & 1 Platoon of K troop led the attack. Ground was completely taken and consolidated by 1315 hrs. 38 Japs killed in this action. Our casualties – 0. During the day, 613 FA got a direct hit on a medium arty peace. One peace thought to be 150 mm. Mortars at 1st & 2d Sqdrn got direct hit on one

70 mm, 1 hvy mort., and one peace thought to be a 77 mm. In addition 2 Hvy & 2 Light machine guns.

Casualties – Approx 100

Lt. Col. Ralph E. Baird, Mars Task Force G-3 filed a report on February 4 that included the following paragraph relative to the 124th:

"Light counter attacks against 2d Sq and 3d Sq positions were repulsed. A Tr and B Tr improved positions by advancing.... G Tr patrolled trails South and West of MONG NOI. F Troop achieved view of road from Milestone 78 to Milestone 75½."

Lt. Col. Baird mention the 124th again on February 7:

"CT 124: No change in positions. Japs patrolled 2 d and 3d Sq positions during night. Seven man Jap patrol walked into L Tr position and was wiped out. Draw North of A and B Tr positions clear of enemy."

Lt. Col. G. T. Laughlin in a report filed on February 9 made the following observations:

"Area Kawngsong, Hpa Pen, Namhpakka clear of Japs. Patrols of I Troop entered last Jap defense position today (Feb. 8) and found 18 dead Japs. This position was an extravaganza of pill boxes, brush and wire. Japs apparently left hastily as they left both the dead and equipment plus American equipment in the positions. 2 Sqdrn found another 4 dead Japs in ravine between them and 1st Sqdrn. Positions unchanged."

The 124th Cavalry moved south along the Burma Road toward Lashio and went into bivouac for training and waiting. They were eventually flown to China to train Chinese army troops. On July 1, 1945 the regiment was disbanded and the men were assigned to other units. In the fall of 1945 they began returning home. For most of the years since they have met in Mineral Wells for a reunion of this unique troop of fighting men.

From John Randolph's Marsmen in Burma:

"On February 18th, Lord Louis Mountbatten visited us. He pointed out that, in the action of all the Allied nations fighting in Burma, the ratio of killed in action was 4 to 1 in favor of the Allies, and that the Marsmen's ratio was 61/2 to 1. He brought tears to the eyes of the cavalrymen as he talked of Jack Knight, of what he had done, and what he represented. And he promised that future maps would show 2nd Squadron's objective hill as "Knight's Hill."

During this visit he also asked Lieutenant General Sultan, his Senior American General to quote him in support of a recommendation that Jack L. Knight be awarded the Congressional Medal of Honor posthumously. This request would help Major Jordan's recommendation for the same award.

The men of Troop F who survived the war returned home and resumed their lives. Although not officially a member of Troop F, Leo Tynan became an attorney and practiced in San Antonio. He has contributed significantly to the details of their battle.

JACK

Jack joined eight of his men from Troop F, another thirteen from other units of Second Squadron and forty-one from the remainder of the 124th Cavalry in their graves on a hillside near the battle scene. The plot was enclosed by a bamboo fence and overlooked a valley of rice paddies. Four years later he was moved to another valley half way around the world to a small cemetery near the family church, Holders Chapel Methodist Church. The day was notable for the bone chilling cold of a mid-winter blue norther that occasionally sweeps through North Texas with gale force winds.

Yet Jack's spirit has not rested through these fifty plus years. We have had periodic reminders of what happened on Loi Kang Ridge during that brutal twenty minutes of February 2, 1945. Following is a list of some of the honors that have come his way.

> The VFW Post in Mineral Wells was named in his honor.
> A street in Fort Wolters was given his name.
> Mamma was named Gold Star Mother of Texas
> The 112th Cavalry National Guard Regiment established a trophy in his name to be presented once each year to a unit for excellent performance.
> On October 14, 1972 Admiral of the Fleet The Earl Mountbatten of Burma dedicated a monument for Troop F, 124th Cavalry in Mineral Wells in which he praised both Jack and Curtis.
> In 1991 a Texas Historical marker was placed at his grave in the Holder's Chapel cemetery.
> In October, 1993 June and I attended a ceremony at Southmost College in Brownsville, Texas to help dedicate a historical marker for Jack who had served at the site of the college when it was Fort Brown.
> In 2000 another historical marker was placed at the entrance to what had been Fort Wolters to honor both Jack and Audey Murphy, both of whom passed through the facility.
> On May 31, 1999 June and I were in Austin to receive on behalf of Jack the Texas Legislative Medal of Honor, the second one to be Presented. Governor George W. Bush made the presentation.

I composed the following poem while attending a superintendents' retreat in the early 1980s. This is my first effort to write about Jack.

JACK

What do you do when a god dies?
Water, ashes, and spark lie,
unable to join in life.
The soul soars to unknown places,
there to join with his God.
I stand within my traces,
bound to this sphere's sod,
praying that it can't be true,
knowing that it is.
Joining with friend and kin in a
wake of paralyzing woe,
We taste the acid of man's sorrow.

A wisp of spirit rises
to shake us loose, our being wakes
with senses and vision of life.
The inexorable flow through the fabric of the soul
sears the pain, contains the woe.
I take up life and memories load
and chart my course, the unknown, with fear.
Small signs of warmth and love
from nurturing lips and hands that knew my fright.
And now, I seek to stand where he stands,
honored, sad, for having him no more.
And his is not yet mine.

CURTIS

Curtis spent four months recovering in hospitals in Burma and India. He arrived home on June 15, the day that Daddy was presented Jack's Medal of Honor. He then spent more time at Brooke Army Medical Center in San Antonio.

On September 10, 1945 Curtis was presented with the Silver Star by Col. William C. McCally, C.O. of Brooke Convalescent Hospital, Brooke Hospital Center, Fort Sam Houston, Texas. The citation read:

"First Sergeant Curtis L. Knight (ASN 20822937), Cavalry, US Army. For gallantry in action on 2 February 1945 near Loi-Kang, Burma. Seeing his troop executive [commanding] officer wounded, and knowing that there were no other officers [of Troop F] available, Sergeant Knight, being the highest ranking non-commissioned officer present, with complete disregard for his own life and under constant enemy small arms fire, assumed the duties of troop executive officer, and led an attack on a strong enemy position, wounded, and unable to move, Sergeant Knight continued to direct the attack from a stretcher, not permitting himself to be evacuated until all other wounded were brought to safety. The courage and bravery of Sergeant Knight reflects great credit upon himself and the Armed forces of the United States."

Curtis was discharged with only ten percent medical disability. He was given a job with Texas Power and Light in the Weatherford office. The son of the president of the company served in the 124th Cavalry. He was Lt. Benjamin H. Carpenter and was awarded a Silver Star for gallantry in action as he fought with another unit of the 124th.

Curtis played baseball for Weatherford in the summer of 1946 in what would be classified now as semi-pro ball. I saw him hit three homeruns in one game against Mineral Wells. He was transferred to Denison, Texas and played baseball for them in 1947. He was offered a job with the Sherman-Denison team in the East Texas League. He thought it was too late for that and soon after went back into the army for a long career.

He spent most of his time as first sergeant of food service schools and on TDY with post baseball teams. I was lucky enough to attend the Fourth Army Tournament in August of 1948 in San Antonio. I got the job of batboy. I was fourteen. Curtis had a good tournament. He batted over .500 and was picked as MVP. He was playing with and against many minor league players who had been drafted into the army.

During 1952-54 he served a three year tour of duty in Japan during which time he was able to reach a state-of -mind toward the Japanese people that erased his racial hatred. The main reason this happened was the gentle and sweet nature of the Japanese lady who worked for them and helped care for Glenda and Larry, his son born the year before.

When he returned he was assigned to the food service school at Fort Hood, Texas. I am sure he was overjoyed to find me in his company. I had signed up for the school during basic training, knowing that it was time for him to rotate and the likelihood that he would be assigned to the school. Everything went well except the morning he chewed us out for dirty latrines and a guy behind me in formation cussed him in a loud whisper. I've always regretted not kicking his ass.

Curtis finished his career at Ft. Hood, eventually taking the job in charge of fish and wildlife conservation and propagation. After he retired from the army, he was able to get the same position authorized under civil service and continued the same job for several years. He cooperated with the Texas Parks and Wildlife Department to conduct research in this well controlled environment.

Curtis has spent much of his time coaching and watching his son, Larry, play baseball and later watching Glenda's sons and grandsons also play baseball. However, Larry's children went another way in sports. Rebecca was good enough in tennis to get a college scholarship and Travis played golf on a scholarship. Larry had pitched baseball for Texas Tech and after being drafted by the New York Yankees, refused to continue a baseball career, choosing instead to work for the City of Killeen in the recreation department.

Curtis lost his dear companion, Joyce, on May 7, 2001 after almost 59 years of marriage. She succumbed to the ravages of diabetes.

LOYD

Loyd finished the war on the island of Pelelieu as a platoon leader and battalion reconnaissance officer. He spent his time there doing some patrol duty, looking for Jap stragglers, playing poker, playing softball and waiting to invade Japan. He built up a good bank account with his winnings. When I said something to him about being lucky, he told me it wasn't luck. He said he played the odds. With his mathematical brain I am sure that he had read a book on the probability in poker and committed it to memory. He could have figured it out for himself. Either possibility is feasible.

When he was discharged, he came home and joined Herchel, R. C. and some of his old Troop F buddies for a wasted year during which all of them blew any savings they had as well as the $20 a week 'rocking chair money' they received from Uncle Sam to help them get re-established. They spent a lot of time in the honky-tonks in Mineral Wells and playing softball for the 'Independents' in the Mineral Wells Commercial League. Curtis also played on this team along with cousin Kenneth Knight and Luther Knight, no kin. They usually had six Knights in the lineup.

After he ran out of money, Loyd tried his hand at construction work in South Texas, working for Scroney Aaron and then as bookkeeper for the same company.

In 1947 he re-enlisted as a technical sergeant but soon applied for his commission and received it. He remarried Lenora in 1950 and was sent to Korea in 1951. While there, he was involved in two traumatic events. One was life threatening and one demonstrates the total absurdity of war. He told Curtis that he was sitting in his foxhole and decided to move to the one next to him to get some information. Immediately after he settled into the second foxhole his first foxhole took a direct hit from a North Korean artillery round. His son, David, told me that on another occasion he was given an assignment to go to the railroad. He took a platoon of men and when they got there, found their job was to pry loose the Korean civilians who were running from the Chinese and hitched a ride on the boxcars. They had frozen to death and stuck to the sides of the cars.

He served in a variety of positions and was sent to Germany in 1955. He was stationed in Munich at SACOM Headquarters. He was assigned to the G-3 Section with his primary duty to plan for the withdrawal of troops and dependents in case of war with Russia.

I was stationed at Neckarsulm with the 9th Ordinance Battalion in support of a battery of 280 mm atomic cannons. I went to see him and exercised the number one prerogative of soldiers and bitched to him about having to do the jobs that three sergeants were doing when I joined the unit. I was a Pfc. Soon after returning to my battalion, I received orders to report to SACOM Headquarters. I was being assigned to Headquarters Company to work in the classified documents section. My office also did work for his office. So I had the distinction of serving in the same company with two of my career military brothers in less than a year.

Loyd's family came over, but David had severe allergies that demanded that Lee take her children home. They were also raising Elaine and Mark Crosland, Lenora's children from her marriage to Micky Crosland. Loyd resigned his commission in a few weeks to avoid spending over two years in Germany without his family. Shortly after returning, however, he re-applied and was reinstated at his rank of captain.

Loyd attended several army schools. He was honor student of an armor school in 1955. He graduated from Air Defense School on July 4, 1957 as honor student. I believe this was the school he was preparing to leave to attend, when two-year old David, who had heard the talk about his going to school, offered some advice. As he was leaving, David advised, "Daddy be careful on the swings."

After serving another tour of duty in Korea in 1960-61 he served the remainder of his time commanding batteries of Nike Missiles in Virginia and battalions around the Dallas-Ft. Worth area. He attended the Command and General Staff College in 1964. He retired May 31, 1967 with the rank of Lt. Colonel.

Loyd died April 7, 1987 at the age of 64. His wife Lee had preceded him in death.

His son, David, has been a high school basketball coach for twenty-five years and has moved to school administration. Elaine Crosland was in the construction supply business and Mark Crosland has a career with the U.S. Postal Service.

HERCHEL

Herchel, after serving on the USS Maryland that took severe hits four times while he was aboard, was discharged in the spring of 1946. He also wasted much of a year until he started dating Lorene Calhoun a member of Roy's Senior Class of 1947. They married soon after her graduation and moved to Washington, D.C. where he worked for a grocery store. They soon returned to Weatherford where he worked in an oil tool factory for a few years. At about the time he started working for General Dynamics in Ft. Worth they had their first child, Jack. Ronnie and Vicki followed.

Herchel would become a machinist, operating a computer controlled milling machine that produced tools and parts for aircraft and missiles. He retired from that job. All three of his children would receive college degrees. Dr. Jack

Knight has a family medical practice in San Antonio. Ronnie was killed in a motorcycle accident and Vickie is a medical technician. His wife, Lorene, died in October 1997.

After Jack returned from Southeast Asia he left active duty in 1979 to return to school, eventually earning his M.D. degree from The University of Texas in San Antonio. He then served as flight surgeon in the Texas Air National Guard. He said his closest brush to tragedy was on a training mission when his F-16 took on a Texas buzzard that flew directly into the engine, causing the plane to crash. He sustained a crushed vertebrae as he ejected. He retired from the military in 1995 but continues in his family medical practice in San Antonio.

ROBERT CLINTON

R. C., who was drafted in May, 1945, was in basic training when the war was over. He later went to Japan to serve in the occupation forces. Mamma managed to harass the army brass until he was discharged in the late spring of 1946.

He joined his brothers in their year of idle play. When that period ended he went to work for Southwestern Bell and was sent to work in the Texas panhandle at Tulia. He couldn't take the isolation. He moved back home, lived with Mamma and Daddy and worked at factories around Mineral Wells until he married Barbara in his early thirties. They had two bright and beautiful daughters, Terry and Karen to go with Barbara's son Mike.

He then was employed by the civil service at Fort Wolters. He stayed there until he discovered he had cancer. He died September 25, 1973, a month short of his 47th birthday.

ROY

I have told much of Roy's story in the text of the book. I have included Gay's story of a little girl who lost her Daddy too soon. I found it very touching as I am sure that you did. Gay now lives in Austin and with her husband, runs a computer business in San Antonio. She has a son named Bryan and a daughter, Jessica. Roy III (Chip) owns in partnership, Bakehouse Cookies, located in Carlsbad, California. He has two sons, Shea and Ryan. Bryan flies for Southwest Airlines.

BILL

I have already mentioned that soon after Jack was killed I formed this notion that somehow, to show my devotion to Jack's memory, that I would think about him at least once every day of my life. It was a shock a few years later that I realized that it had been days or perhaps weeks since I had given him any thought. I had become too involved with my own life to waste time on obsessions. Since my retirement, I rediscovered the obsession.

I was the fortunate recipient of a lot of positive regard, much of it undeserved. The result of this was that I actually thought that I was supposed to excel at the things I did.

I kept plugging along, going to college, spending two obligatory years in the army. I had little natural inclination toward warrioring. I received four degrees: the AA from Weatherford College, B.S and M.S. from North Texas State University in education, and a Ph.D. in school administration from East Texas State University. (Now Texas A & M, Commerce)

I married Shirlene Leach, a great, bright and beautiful little blonde who thought she could do anything I could do. Well she did, except her last degree is designated Ed. D in Special Education. I taught, was a principal and spent 27 years as a school superintendent. Shirlene has taught 30 years, getting a late start after serving as my blonde scholarship.

We have three sons, Matt, Scott and Joe. Scott has a M.S. degree in school administration and is a school principal. Joe has a B.A. degree in English and is a graduate of the National Shakespeare Conservatory of New York City. He has worked in management at Arena Stage Theatre in Washington, D.C. and is now doing free-lance work in a number of endeavors. He is married to Andrea Maida, an actress and a highly talented, delightful lady. Matt is a heck of a golfer and a great lawn care specialist. He is also one of the bravest people I have ever known, having battled his schizophrenia-induced demons for more than twenty years.

Scott and his wife Denice, also a teacher, have given us three beautiful and bright granddaughters, Becca, Bethanne and Hannah. Shirlene spoils them. I just love them.

JUNE

Baby Sister and I are great friends. She is absolutely one of the best people on earth. She finished high school in 1954 and went to a business college for a year. She then entered the civil service and had a forty-year career, most of which was as a transportation officer, responsible for moving personal property, DOD passengers and dependents, and cargo for military personnel and government contractors. She had a rating of GS-12 the last several years. She could have been promoted, but she was not willing to put her career ahead of her family.

June and her husband, Wayne 'Soup' Campbell have two children, Wayne 'Bubber' and Melody. Bubber works for a construction company as a supervisor for the installation of elevators in high-rise buildings. Melody is a teacher's aide.

A few years ago June and Soup bought the old Knight family farm at the bend of a creek in a beautiful setting just under the rim of a rock cliff. June is retired but still plays the piano for her church and dotes on her grandchildren and great grandchildren. After a career in public schools as a school principal, Soup occasionally goes overseas to supervise workers for contractors employed by the U.S. Government to build facilities. He has worked in Kosovo and Afghanistan.

MAMMA and DADDY

Daddy bought a sixty-acre farm with Jack's savings, as was his wish. It was located one mile north of Millsap. The farmhouse was painted white with a grass yard and white picket fence. We resumed a life style much like we had before the war, except we were the landowners. We would soon have electricity and butane gas for heat. A few years later Millsap would put in a water system and our farm was the first customer on the line. We were then able to have a bathroom.

Old habits we found hard to break or maybe Daddy just didn't want to deprive June and me of the fun of picking cotton for other farmers and harvesting pecans on the creek bottoms. He kept that up into the sixties. I recall taking Matt, my oldest son along on one of these workdays. He loved it. He was three and was very proud when he filled his one-pound coffee can with pecans.

They never relented in their Christian commitment. Mamma would eventually follow her heart and join a Pentecostal church, The Assembly of God, along with June. Daddy soon followed. I am the only member of the family who has maintained the family Methodist tradition.

They were always affected by the military, having three sons making careers there. When one would go overseas, their wives and children would live with Mamma and Daddy until they joined their husbands. As a result of this, my brothers had a new, more conveniently arranged, house built for them on the farm in 1957.

Both of our parents inherited traits that made them susceptible to circulatory health problems. Their rich diets of farm foods with too much prohibitive fats caused both to have early problems. Daddy had a stroke in his early sixties, but survived to work again. In his early seventies, Daddy's mind would succumb to the ravages of arteriosclerosis and he became comatose the last two years of his life. Mamma had a severe stroke the week after Christmas 1969 and died January 14, 1970. Daddy lived another thirteen months, dying February 14, 1971.

BRUNO

Bruno lived in semi-retirement after being shot while killing goats. His only combat experience after that was to half-heartedly chase a chicken out of the yard. He died in 1955 of old age while I was on my tour of duty in Germany. He was a good dog except for his felonious conduct with goats.

After all the years of thought and tears, research and writing, the melting pot of my mind and soul created this, my final thoughts on the effects of war.

TRAGEDY

It is a tragedy that Jack died, but the real tragedy was that he didn't grow older to see June grow up, enjoy Mamma's and Daddy's having a time free of debt, that he didn't get to show he could have true love for his little Mexican beauty. Tragically, he didn't have time to complete his growth toward the realization that black people and the Japanese as a people are worthy of respect, that he couldn't use his courage, talents and energy to create.

It is a tragedy that Roy couldn't complete his career and re-join his dear family to grow old with Patricia and see Chip, Gay and Bryan fulfill their great promise as bright and talented young adults. It is a tragedy that their extended families had both their presence and influence no more.

The tragedy of war for all time is that men of good will and gentle nature must diminish their souls and minds so that they must, within the same breast, harbor a love and compassion for their fellow warriors while fomenting hate for fellow human beings, who also loved and hated, to satisfy the need to retrieve equilibrium in the political world.

After the War

Curtis receiving MVP Trophy after Fourth Army Tournament, 1948

Bill in Germany, 1955

Herchel, R.C., H.D., Curtis

Between Wars, 1952
R.C., Bill, Herchel, Roy, Jr., Curtis, Larry & Jackie

ACKNOWLEDGEMENTS

I COULD NOT HAVE written this book without substantial help. I must first credit my wife, Shirlene, for her encouragement, advice and patience. Our son, Scott and his wife Denice, lent a hand. The most extensive hands-on assistance came from Joe, another son, the English major. Joe provided technical assistance, research and invaluable advice in all aspects of the project. His wife, Andrea, also added her insightful advice, as did Scott Burgess and Maia DeSanti. To Matt, my oldest son, thanks for a model of courage to help me press on. Our granddaughter, Becca helped create the maps.

My sister, June and niece, Gayann provided recollections in writing that are included in the text. My brother, Herchel told me of his experiences on board the U.S.S. Maryland during World War II. Nephews Bryan, Roy III, David, and Jack helped by relating their own stories and by providing encouragement and historical records. My niece, Elaine, provided historical material and contacts with Troop F veterans. I simply could not and would not have completed this work without the approval, encouragement, verification, validation, and vital recollections that Curtis so willingly and painfully provided. His dear wife, Joyce, before her death, contributed her colorful and romantic memories. Thank you, Joyce, for making all our lives infinitely richer and more interesting.

Special thanks must go to the few warriors of Troop F who hung on to life long enough and who were willing to submit to my intrusive interviews. This brave and congenial group included Bill McQuary, Mickey Crosland, Doc Clabuesch, Pressley Hinson, Charles Baker, W.T. Lowthorp, Alvin Moberly, Kenneth Payton (with son-in-law Robert Orr, O.D.), Hugh Warren, John Warren, Bea Alco and Jack Warner. Others at the Battle of Knight's Hill whose recollections added substance to the account were Second Squadron Surgeon, J. M. Huey, M.D. and Leo Tynan who so courageously took over command of Troop F after all its officers were killed or wounded. David Lee, son of F Trooper Ambrose Lee, has been helpful by allowing me time at their reunions to speak and interview. To the wives of these veterans, thanks.

I absolutely must thank John Randolph posthumously for his detailed account of the Mars Task Force in his book Marsmen in Burma.

Two ladies deserve my special gratitude. Miss Lou Sullivan and Miss Cleo Smith were Jack and Curtis's teachers in elementary school. Into their nineties they still had clear memories that confirmed the early character of these brave men. Their frank and insightful comments are appreciated.

Ken Slesinger of the National Archives II in Maryland and the staff were exceedingly kind and helpful during my three days at their facility and after. Tom Watts, my library friend and Angela Grigson, our British friend and professional editors, did what they could. Bonnie Mathis, a friend of our friend JoAnne McMillon, also experienced in editing was extremely helpful in locating minute errors and suggesting other changes that proved helpful.

Many other friends read and contributed to the final product. Our intellectual friend Norma Bruce, gave me the first glimmer of hope that I might have had something more than a high school term paper. Thanks go to Jerry Cloud and Cynthia Clark, also professional colleagues, for reading and commenting on the text. Larry Lewis, another friend and colleague deserves my gratitude for his enthusiastic endorsement of the book and most of all, for suggesting the title. To Col. Jack Russell (ret.), friend, colleague, warrior, patriot for reading and commentary, thanks. Laura Simms provided much needed expertise in dealing with that confounded computer. Larry Arnold has my gratitude for his keen interest in our family and encouragement for this project.

Being an educator, I am professionally and spiritually obligated to honor my profession. I have always remembered all my teachers with fondness and credit them for helping guide me through life. In the context of this endeavor, however, I will give preeminent credit to three English teachers and a university professor. Thanks Yvonne Huffstutler, Billie Bell and Jack Harvey for insisting that I write a coherent sentence and a passable paragraph. Thank you Dr. Lynn Turner for demanding logic, accuracy, cogency and attention to detail during the writing of my dissertation. All those lessons made this possible. If this effort succeeds, you share much of the credit. If it fails, the failure is all mine.

BIBLIOGRAPHY

From Basic to Burma, Joyce Knight, Editor, Unpublished.

History of Parker County, Parker County Historical Commission, Taylor Publishing Company, 1980.

Mars Task Force, World War II, Operations Report, Entry 427, Record of the Adjutant General's Office (R.G. 407), National Archives at College park, Md.

Marsmen in Burma, John Randolph, Gulf Publishing Co. and Curators of the University of Missouri, 1946, 1977.

New History of World War II, C.H. Sulzberger with revision by Stephen E. Ambrose, American Heritage, Viking Penguin, 1966, 1994,1997.

U.S.S. Maryland Deck Log, World War II, Records of the Bureau of Naval Personnel (R.G. 24), National Archives at College Park, Md.

'Theme', Jack Knight, 1937, unpublished.

Family letters, unpublished.

Knight Medals
A Key to the Back Cover

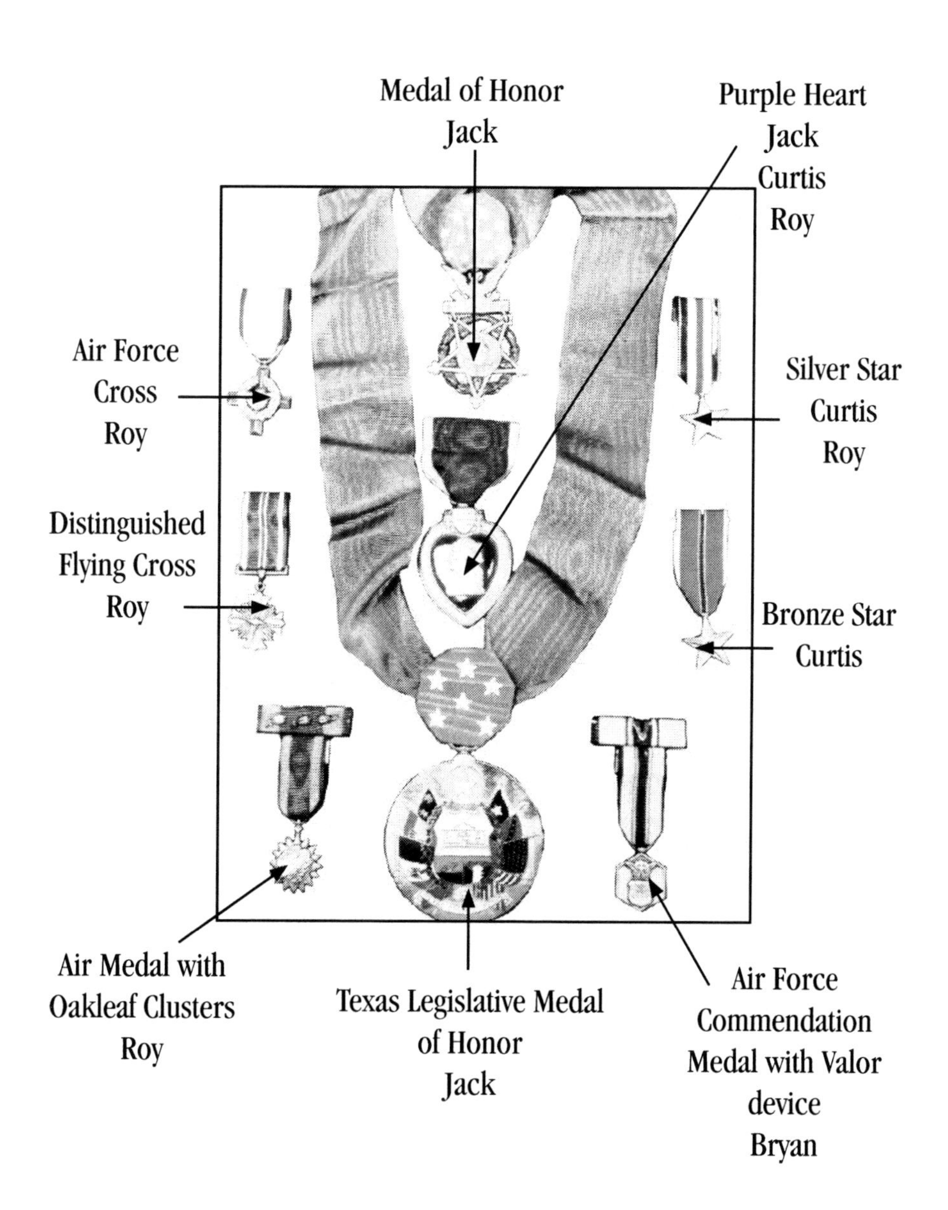

ISBN 141203692-5

9 781412 036924